TECHNICAL ARITHMETIC AND MENSURATION.

By CHARLES W. MERRIFIELD. F.R.S. Barrister-at-Law, Principal of the Royal School of Naval Architecture and Marine Engineering. Honorary Secretary of the Institution of Naval Architects, and Late an Examiner in the Department of Public Education; Joint-Editor of the Series. Price 3*s*. 6*d*.

KEY TO MERRIFIELD'S TECHNICAL ARITHMETIC AND MENSURATION.

By the Rev. JOHN HUNTER, M.A. one of the National Society's Examiners of Middle-Class Schools; formerly Vice-Principal of the National Society's Training College, Battersea. Price 3*s*. 6*d*.

ON THE STRENGTH OF MATERIALS AND STRUCTURES:

The Strength of Materials as depending on their quality and as ascertained by Testing Apparatus. The Strength of Structures, as depending on their form and arrangement, and on the materials of which they are composed. By JOHN ANDERSON, C.E. LL.D. F.R.S.E. Superintendent of Machinery to the War Department. Price 3*s*. 6*d*.

ELECTRICITY AND MAGNETISM.

By FLEEMING JENKIN, F.R.SS. L. & E. Professor of Engineering in the University of Edinburgh. Price 3*s*. 6*d*.

WORKSHOP APPLIANCES.

Including Descriptions of the Gauging and Measuring Instruments, the Han Cutting-Tools, Lathes, Drilling, Planing, and other Machine-Tools used by Engineers.
By C. P. B. SHELLEY. Civil Engineer, Honorary Fellow and Professor of Manufacturing Art and Machinery, King's College, London. With 209 Figures on Wood. Price 3*s*. 6*d*.

The following Text-Books *are in active preparation:—*

ORGANIC CHEMISTRY.

By H. E. ARMSTRONG, Ph.D. Professor of Chemistry in the London Institution.

PRACTICAL AND DESCRIPTIVE GEOMETRY, AND PRINCIPLES OF MECHANICAL DRAWING.

By C. W. MERRIFIELD, F.R.S. Principal of the Royal School of Naval Architecture, South Kensington; Joint-Editor of the Series.

A MANUAL OF QUANTITATIVE ANALYSIS.

By T. E. THORPE, F.R.S.E. Professor of Chemistry in the Andersonian University Glasgow.

MANUAL OF QUALITATIVE ANALYSIS AND LABORATORY PRACTICE.

By T. E. THORPE, F.R.S.E. Professor of Chemistry in the Andersonian University Glasgow; and M.M. PATTISON MUIR.

PRINCIPLES OF MECHANICS.

By T. M. GOODEVE, M.A. Lecturer on Mechanics at the Royal School of Mines, and formerly Professor of Natural Philosophy in King's College, London; Joint-Editor of the Series.

ELEMENTS OF MACHINE DESIGN.

With Rules and Tables for Designing and Drawing the Details of Machinery. Adapted to the use of Mechanical Draughtsmen and Teachers of Machine Drawing.
By W. CAWTHORNE UNWIN, B.Sc. Assoc. Inst. C.E.

ECONOMICAL APPLICATIONS OF HEAT.

Including Combustion, Evaporation, Furnaces, Flues, and Boilers.
By C. P. B. SHELLEY, Civil Engineer, and Professor of Manufacturing Art and Machinery at King's College, London.
With a Chapter on the Probable Future Development of the Science of Heat, by C. WILLIAM SIEMENS, F.R.S.

THE STEAM ENGINE.

By T. M. GOODEVE, M.A. Lecturer on Mechanics at the Royal School of Mines, and formerly Professor of Natural Philosophy in King's College, London; Joint-Editor of the Series.

SOUND AND LIGHT.

By G. G. STOKES, M.A. D.C.L. Fellow of Pembroke College, Cambridge; Lucasian Professor of Mathematics in the University of Cambridge; and Secretary to the Royal Society.

*** *To be followed by other works on other branches of Science.*

London: LONGMANS and CO. Paternoster Row.

TEXT-BOOKS OF SCIENCE

ADAPTED FOR THE USE OF

ARTISANS AND STUDENTS IN PUBLIC AND OTHER SCHOOLS

WORKSHOP APPLIANCES

WORKSHOP APPLIANCES

INCLUDING

DESCRIPTIONS OF THE
GAUGING AND MEASURING INSTRUMENTS,
THE HAND CUTTING-TOOLS, LATHES, DRILLING,
PLANING, AND OTHER MACHINE-TOOLS
USED BY ENGINEERS

BY

C. P. B. SHELLEY

CIVIL ENGINEER,

Honorary Fellow of, and Professor of Manufacturing Art and Machinery in King's College, London

D. APPLETON AND CO.
549 & 551 BROADWAY
NEW YORK
1873

PREFACE.

THIS TREATISE is intended to give, in a concise form, an explanation of some of the contrivances used in the Engineer's Workshop. The recent establishment of Scholarships, to which the name of Sir JOSEPH WHITWORTH, Bart., is so nobly attached, having, it is thought, rendered the present work necessary as an aid to those who intend to offer themselves as Candidates for this honourable distinction.

The student, in making himself familiar with the appliances described, will but be following in the steps of many of our most eminent civil engineers, some of whom have made improvements in mechanism, upon which, however trivial they may appear in themselves, has depended the success of the most brilliant of their achievements.

To the workman, many of the explanations which the book contains may indeed be superfluous; but the ever-increasing demands for greater accuracy in his handicraft render it very desirable that he should have the means of acquainting himself with the present methods of obtaining delicate measurements, a subject upon which it has hitherto been difficult for him to obtain information.

The thanks of the Author are due to Sir Joseph Whitworth; also to Messrs. Sharp, Stewart, & Co., of Manchester; to Messrs. Shepherd, Hill, & Co., and to Messrs. Fairbairn, Kennedy, & Co., of Leeds; to Messrs. Tangye Brothers, of London, and to Mr. J. J. Bagshawe, of the Thames Steel Works, Sheffield, for their liberality in supplying photographs &c., from which many of our illustrations are taken. He is also greatly indebted to his friend Mr. A. L. Newdigate, for so kindly rendering his valuable assistance and co-operation during the progress of the work.

113 Victoria Street, Westminster.

CONTENTS.

CHAPTER I.

ON MEASURES OF LENGTH AND METHODS OF MEASURING.

CHAPTER II.

ON HAND TOOLS USED FOR WOOD.

CHAPTER III.

ON HAND TOOLS USED FOR METAL.

CHAPTER VIII.

ON PLANING, SHAPING, AND SLOTTING MACHINERY.

CHAPTER IX.

ON PUNCHING AND SHEARING MACHINERY.

CHAPTER X.

ON THE DISTRIBUTION OF THE MOTIVE POWER TO MACHINE-TOOLS.

LIST OF TABLES.

WORKSHOP APPLIANCES.

CHAPTER I.

ON MEASURES OF LENGTH, AND METHODS OF MEASURING.

MORE than two hundred and fifty years ago the greatest English writer on philosophical subjects, wishing 'to quicken the industry and rouse and kindle the zeal of others,' expressed himself as follows :—'The introduction of famous discoveries appears to hold by far the first place among human actions; and this was the judgment of former ages. For to the authors of inventions they awarded divine honours, while to those who did good service in the State (such as founders of cities and empires, legislators, saviours of their country from long endured quarrels, quellers of tyrannies, and the like), they decreed no higher honours than heroic. And certainly if a man rightly compare the two, he will find that this judgment of antiquity was just. For the benefits of discoveries may extend to the whole race of man—civil benefits only to particular places; the latter last not beyond a few ages, the former through all time. Moreover, the reformation of a State in civil matters is seldom brought in without violence and confusion : but discoveries carry blessings with them, and confer benefits, without causing harm or sorrow to any.'

With respect, however, to the motives of the inventors to whom these high honours should be accorded, Lord Bacon adds : 'Further, it will not be amiss to distinguish the three kinds, and as it were grades, of ambition in mankind. The first is of those who desire to extend their own power in their native country; which kind is vulgar and degenerate. The second is of those who labour to extend the power of their country and its dominion among men. This certainly has more dignity, though not less covetousness. But if a man endeavour to establish and extend the power and dominion of the human race over the universe, his ambition (if ambition it can be called) is without doubt both a more wholesome thing and a more noble than the other two. Now the empire of man over things depends wholly on the arts and sciences. For we cannot command nature except by obeying her.'*

Rarely, indeed, are our inventors actuated by this highest form of ambition at the present day; but, whatever may have been their motives, we cannot deny that they have been in a high degree successful in extending the power of their country. Where, for example, should we now be in the application of steam-power, if the capabilities of our workshops did not exceed those of the days of Newcomen—or even of Watt? Where would be our naval strength, or even our ability to defend our shores, if the increased size and precision of our artillery had not been rendered possible by the previous introduction of the steam-hammer and the power-lathe? So that, without in the least desiring to detract from the fame of the individuals to whom we owe the inventions, or the series of inventions, by which each crude idea has been gradually brought to comparative perfection, we cannot but consider that their fellow-labourers —who, by the introduction of the appliances known as 'machine tools,' &c., have rendered feasible the carrying out

* *Novum Organum*, lib. i. aph. cxxix. Quoted from Ellis and Spedding's translation.

of their suggestions, and the performance of operations which before were utterly impracticable—have strong claims to a share in the credit.

Under this aspect, then, the workshop appliances which form the principal subject of the present volume will be seen to possess more than a mere technical interest, inasmuch as the mechanical superiority which lies at the root of our commercial prosperity, is to a great extent dependent upon them. It may be worth while, therefore, to pause for a moment, and consider in the first place whether we can assign their present state of efficiency to any particular cause or causes; and secondly, whether any one direction seems to be more marked out than another for future progress—for who will say that there is no room for further improvement, or question the necessity for continued efforts, if we desire to maintain the lead in competition with our neighbours?

With regard then, in the first place, to the sources and the limits of our present powers in the execution of metal work, we shall perhaps best appreciate our own position in this respect by calling to mind the obstacles which hampered the progress of the early millwrights. These were chiefly the want of appliances for treating any large portions of work, on account of their mere size, and the inability to produce true cylindrical or flat surfaces even in work of moderate dimensions. Of these, the former has now altogether ceased to be regarded as an impediment, the steam-hammer enabling us to forge, and our Herculean power-lathes and other machines rendering it an easy operation to give the requisite accuracy of form to the very largest masses which have as yet been required in the arts of either peace or war. Nor indeed does the limit in this direction seem to be otherwise than capable of indefinite extension. For the accurate performance, and indeed for the existence of our machine-tools, both large and small, and consequently for our enormously increased facilities for giving to work of all sizes the requisite truth of form, we are indebted to the invention of the slide-rest and the planing-machine;—or rather to the introduction of the *principle of the slide*, upon which they both depend. The application of this principle, then, seems to have been the turning-point in our mechanical career, from which the rapid strides which we have made of late years may be said to have commenced. Not only has it given us the various forms of self-acting lathe, the planing, shaping, slotting, and other machines which have become absolute necessities in our workshops, but, above all, it has made it possible to increase the dimensions of our work—by increasing the size of the tools in which it is treated—to an extent to which at present we have found no limit.

The slide principle consists merely in so forming two adjacent plates or other portions of a machine, that one of them can slide freely upon the other in one direction only—no other motion between them being possible. Can anything be more simple? Yet simple though it be in conception, and various as are the forms of sliding-joint by which these requirements may be fulfilled, its embodiment in a practical shape demands a very high degree of manipulative skill, as the reader will have no difficulty in realising when, for instance, the slide-rest is placed before him. Here he will see a structure, formed by building up a series of slides, one upon another, each of which must be incapable of receiving any except its own proper motion; since the slightest unsteadiness would be communicated to the tool supported on its summit, and would render the whole instrument useless. The greatest care therefore becomes necessary to ensure perfect parallelism and exact agreement between the surfaces which are in contact. So great indeed does this necessity become in the larger and more powerful tools, that it is probable that their production would have continued to be beyond our powers even to the present day, but for the improved process first introduced to public notice by Sir Joseph Whitworth in the year 1840—by which the formation of true plane surfaces has been rendered possible.

Slight though the connection may at first sight appear to be, we believe that this improvement provides us with the most probable answer to our second enquiry. Since the true plane (or at least a very close approximation to it) has become a reality, increased accuracy has gradually extended to all branches of the mechanical arts in which true surfaces are required. This is of itself an advantage, which has been, and will undoubtedly continue to be, attended with increasing benefits; but its own importance will in all likelihood prove secondary to that to be derived from the admirable system of measurement which has more recently been devised and brought to a high state of perfection by the same master of perseverance.

To the success of this principle,—which consists in, as it were, magnifying minute quantities by mechanical instead of by optical means, as explained hereafter,—the production of a true plane has to so great an extent conduced, that without it there is no doubt that its application would have been impracticable. As it is, the measuring machines in question constitute, in their most delicate form, the most beautiful mechanical appliance of our time, and are of the highest value for scientific purposes,—whilst those of more humble powers supply us with the means of obtaining measurements with an amount of precision which at present far exceeds our workshop requirements. That these requirements will advance as the benefits derivable from increased attention to dimensions and the abandonment of haphazard fitting—that insuperable barrier to the economical production of a high class of work—become more and more fully appreciated, is our firm belief; and we look forward to seeing the day when some kind of efficient measuring instrument, by which his gauges can be checked and the constant accuracy of his dimensions can be ensured, will have become a necessary part of the stock-in-trade of the machinist.

In view, then, of the great and increasing importance of accurate measurement, we propose to devote the present chapter to noticing some of the measuring instruments which are now in use in mechanical workshops, together with a few of the more delicate appliances which may in some cases be substituted for them with advantage; or in others be employed for their correction and adjustment.

The necessity for some *Standard*, or Unit of Length, appears to have been recognised from the earliest times; but until the seventeenth century—at about the commencement of which some demand for increased accuracy seems to have sprung up—various natural objects served the purpose sufficiently well. Many of the names which survive to the present day show the origin of the measures to which they are applied. Thus we use the 'foot,' the 'hand,' and the 'nail' for the expression of various small dimensions; whilst the 'cubit' (or forearm) and the 'pace'—though the former is not to be met with now—have in past times served to express somewhat greater ones. It was doubtless an advantage to make use of units, with duplicates of which every workman or trader was provided; but inasmuch as they all vary to a great extent in different individuals, they could obviously only be used for measurements of the roughest description. Attempts were made long ago in this country to secure one permanent unit, or standard of length, and although they have been successful as far as commercial and other practical ends are concerned, the history of our endeavours to obtain one sufficiently invariable for scientific purposes is a standing example of the uncertainty of human undertakings.

Some of the old statutes enact that three barley-corns, round and dry, make an inch, twelve inches a foot, three feet a yard, &c., and there seems to be no doubt that this mode of obtaining the standard was actually resorted to. But, setting aside the objection due to the varying sizes of individual grains—unless the average from a very large number of them be taken—it is so difficult to know how much of the sharp end of a grain of barley must be removed in order to make it 'round,' that the definition is not of much value. Nevertheless, in spite of numerous attempts at legislation on the subject, this, down to the year 1824, was the only process by which the standard yard of this country could, if lost, be legally recovered.

In the year 1742, when the Royal Society took the matter in hand, there were several standards of length in existence, which, although they were not recognised by law, served the purpose of approximately regulating the measures used for commercial and other purposes. A careful comparison made by order of the Royal Society showed considerable variation between them; so from the most reliable of them a clever mechanician named Graham (to whom we owe the mercurial pendulum, and several other improvements in clocks and watches) was employed by that Society to mark off the true length of a yard upon a brass rod. From this —the first standard worthy of the name—a Parliamentary Committee caused a copy to be made in the year 1758 by an optician named Bird, who was also afterwards employed to make a second copy. These two standards were duly placed in the hands of the Speaker of the House of Commons, but a Bill to constitute the first of them the national standard was not carried; so there they remained. Another Committee appointed in 1790 did nothing, and the matter was then shelved till the year 1814.

During this interval, however, not only did several private individuals and learned societies provide themselves with standards, but scientific men turned their attention to devising means for the recovery of the national standard—if it should ever be lost or destroyed. Accordingly a Committee in 1814, after recommending the adoption of Bird's standards, defined the British yard as consisting of thirty-six inches,—the length of a pendulum making one vibration per second being 39·13047 of such inches. In 1819 a Commission was appointed by the Prince Regent, which in that and the two following years presented three Reports. In the first Report the length of the seconds' pendulum, and also the weight of a cubic inch of distilled water, were stated; but in the second and third Reports respectively both of them were altered,—the former to 39·13929 inches, the latter to 252·458 grains,—under certain stated conditions. These Reports resulted in the Act of 1824, by which Bird's standard of 1760 was at length legalised.

Only ten years after this, in 1834, the fire occurred which destroyed the Houses of Parliament, and with them both of Bird's standards; so the country again found itself without an authorised copy of its standard of length. It might naturally be supposed that recourse would at once be had to the pendulum, and that the easy and speedy restoration of the lost standard would be the result. But this was not the case; and a Treasury Commission appointed in 1838 declined to recommend that method, serious sources of error in the former experiments having been subsequently discovered. Fortunately, however, the Royal Astronomical Society had, only two years before the destruction of the Houses of Parliament, caused a standard to be made for its own use, and this had been most carefully compared with the Parliamentary standards. From it, therefore, taken in conjunction with other reliable copies, the Commissioners (whose Report was presented in 1841) recommended that accurate Parliamentary copies be made, by the careful preservation of which the true length of the British yard should be perpetuated, in preference to depending upon experiments for obtaining it from any natural constants whatsoever.

Acting on this Report, the Government appointed a Committee to carry out the recommendations contained in it, under which was commenced a series of experiments and investigations which are almost, if not altogether, without parallel for the care and perseverance bestowed upon them.

Those connected with the restoration of the standard of length—with which alone we have to deal—were begun by Mr. Baily (to whose assiduity the Astronomical Society owed its standard), and after his death in 1844 were carried on by Mr. Sheepshanks. The delicacy of their operations may be gathered from the fact that a change of temperature of only one-hundredth part of one of Fahrenheit's degrees was found to produce a sensible effect in the length of a bar; a quantity which at that time no thermometers in the country were capable of registering. Therefore, in addition to accurate determinations of the rate of expansion in different metals, careful comparisons of the most trustworthy copies of the standard, and many other points to which time had to be devoted, new thermometers of immensely increased accuracy had to be devised and constructed. To this arduous task Mr. Sheepshanks gratuitously devoted the last eleven years of his life, dying on the very eve of its completion, in 1855. What remained to be done was undertaken by Mr. (now Sir George) Airy, the present Astronomer Royal, to whose Memoir in the 'Philosophical Transactions' for 1857 the reader who is interested in the subject will do well to refer for further details of the work.

The result of their joint labours was the adoption of two of the existing copies of the lost standard, one of which was henceforward to be regarded as *the* standard. From this four 'Parliamentary copies' were made, of which No. 1 was deposited in the Royal Mint, No. 2 in the keeping of the Royal Society, No. 3 in the Royal Observatory at Greenwich, and No. 4 was immured in the lower waiting hall of the Houses of Parliament.* In form they are all

* Recent alterations at the Houses of Parliament have rendered necessary the removal of this Parliamentary copy, together with those of the Standards of Weight, &c., which were immured with it; the wall in which they were originally placed having been pulled down in order to form an entrance to the refreshment rooms. The *Times* (of March 21st, 1872) states that:—'On the 7th of March . . . the

solid bronze bars, 1 inch square, each being 38 inches long. Near the ends of each bar there are two holes, in which are inserted gold plugs with fine lines engraved upon their surfaces, which lines are exactly one yard apart, at a certain stated temperature. An Act legalising them was passed in 1855, since which date—despite the efforts which have been made in this country to implant the mètre with its decimal system—these bars have continued to be the only legal source from which all our measures spring, and the real safeguard against the loss of our national unit, the British yard.

In addition to the length of the seconds' pendulum, other natural standards have from time to time been proposed. Such are, the velocity of light, the length of an undulation of a ray of light of definite refrangibility (neither of which are practically available), and various dimensions of the earth itself. One of these has been adopted by the French, whose unit of length, the mètre, professes to be one ten-millionth part of a line drawn on the surface of the globe from the pole to the equator. The length of this line was deduced from measurements of the arc of the meridian between Dunkirk and Barcelona. The result of these measurements, which were commenced in 1792, has since been found to be incorrect by an appreciable quantity, and the length of the mètre, if once lost, could with no greater certainty be recovered from any natural standard than can the British yard. Consequently the decree passed by the French Republic in 1795 wisely defined the mètre to be the distance between the ends of a standard bar of platinum, without reference to the source from which it had been derived. Borda determined the exact relation which the length of this bar bears to that of a pendulum vibrating seconds in the latitude of Paris, but the true safeguard against the loss of

standards were deposited in their new resting-place in the wall on the right hand side of the second landing of the public staircase, leading from the lower waiting-hall up to the Commons' Committee rooms.'

the French, as of the English standard, lies in the preservation of accurate copies of the bar by which it is represented.

The length of the standard mètre is 39·37079 British inches, which for practical purposes may be easily remembered as *three feet three inches and three eighths of an inch*, this being only ·00421 (about $\frac{1}{240}$) of an inch in excess, which is a quantity quite inappreciable in any ordinary measurements. Some further information concerning the French standard, and the connection which exists between the units of length, capacity, and weight in the metric system, will be found in the volume of this series on 'Heat,' by Professor J. Clerk Maxwell, and also in that on 'Inorganic Chemistry,' by the late Doctor W. A. Miller. Where large numbers of weights or measurements have to be converted from the metric to the British system, or vice versâ, time and trouble may be saved by laying down contiguous scales of equal parts, bearing the proper proportion to each other —on the principle of the 'Metric Scales,' by Mr. A. L. Newdigate, recently published.

It has been proposed by Sir John Herschel to adopt the earth's polar axis as a fundamental unit of length. For this purpose it might be considered to contain 500,500,000 (five hundred million five hundred thousand) British inches, which is not more than 82 yards (about $\frac{1}{170000}$) in excess of its true length.

In the restoration of the standard yard, the means employed for detecting the minute differences of length which had to be dealt with, consisted of a pair of microscopes attached to a perfectly firm foundation, in the focus of each of which were placed two intersecting cross-hairs. A slow traversing motion could be given by means of a micrometer screw, the head of which was so divided that each graduation represented a movement of one twenty-thousandth part of an inch. The comparison of two standard or other scales was effected by accurately adjusting the points of intersection of the cross-hairs over the lines engraved upon one of the scales, then replacing it by the second scale and again adjusting the micrometer screw, the graduations on the head of which registered the difference in the length of the scales.

This method is known as 'line' measure, or *mesure à traits*, in contradistinction to *mesure à bouts*, or 'end' measure. In the former, as we have just stated, the length is given by the distance between two fine transverse lines drawn upon the standard bar, and it is read by the aid of micrometer microscopes; in the latter it is given by the distance between the ends of the bar itself, which is placed in actual contact with the instrument by which it is measured. Sir Joseph Whitworth has shown that for this purpose the sense of touch is much more reliable than the sense of sight, inasmuch as minute distances can be enlarged with much greater certainty by mechanical than by optical means. On this principle he has constructed machines capable of detecting a difference of one-millionth part of an inch in the length of two bars; and he considers that with the assistance of these there would be no difficulty in producing any required number of copies of the Parliamentary Standard, each one of which would be quite sufficiently accurate for all ordinary purposes of reference, so that the wear and risk of injury to the original standard would be reduced to a minimum.

Through the courtesy of the inventor we are able to give an engraving of one of these machines, the explanation of which cannot fail to interest the reader. It must be borne in mind that its object is not to make an original measurement of the total length of any bar, but to compare it in the most accurate manner possible with a nearly similar standard bar of which the exact length is known, and record, to the millionth part of an inch, any difference which may exist between them. The particular machine here represented is constructed to receive a bar only one inch long; but one made upon a precisely similar principle, which was capable of taking in a bar forty inches in length, was shown in the Great Exhibition of 1851, and obtained for Sir Joseph Whitworth the award of the Council Medal. Another machine for 'end' measurement was also exhibited there by the Prussian astronomer Bessel.

The appearance of the instrument will be evident from the General View which stands at the head of the following page, whilst the outline Plan, Section, and End Elevation below will explain its construction—the same parts being indicated by the same reference letters in all of them. A standard one-inch bar (D) is there shown in position for being measured—or rather for being compared with some other bar. A rigid mass of cast iron (A) forms the bed of the machine, the casting being carried up at each end so as to form two head-stocks. Running from one of these headstocks to the other is a V-shaped groove, in which the square bars B and C are laid, and which also receives the standard or other bar (D), of which the length is to be tested. The sides of the groove, and also those of the bars (which are square in section), are worked up as truly plane as possible, and are kept accurately at right angles to each other, so that, upon whichever sides the bars may rest, they are capable of sliding smoothly and with perfect steadiness in the groove. Their ends also are carefully made square to their sides, and are brought to true planes; one extremity of each, in the case of B and C, and both extremities of D, being turned down, so as to present circular instead of square faces. Through each head-stock runs an accurately pitched micrometer screw, by which B and C can be driven forwards along the groove; as may be observed in the left-hand portion of the Plan, in which the saddle by which B is protected and partially concealed, when the machine is in use, has been removed. The screw on this side, which has exactly 20 threads to the inch, is made to advance or recede by turning the wheel F, the circumference of which is divided into 250 parts. Consequently, by turning the wheel forwards through one division, the bar B is moved through $\frac{1}{20} \times \frac{1}{250}$ = one five-thousandth of an

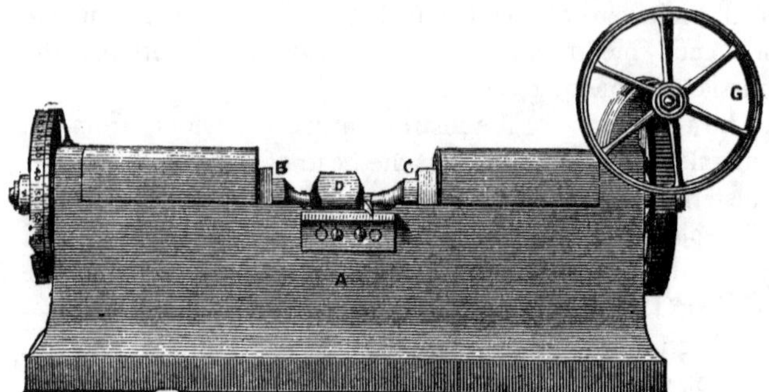

GENERAL VIEW.

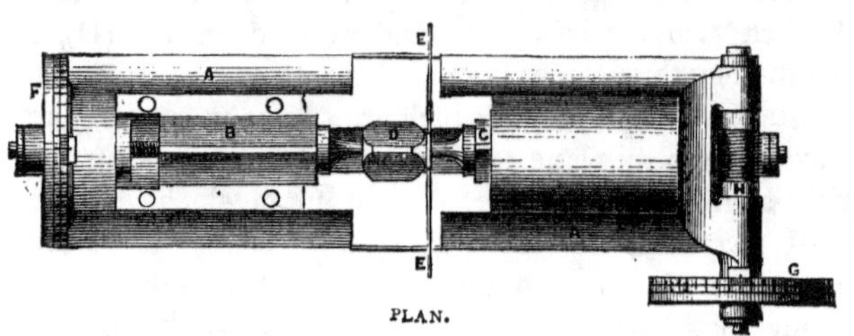

PLAN.

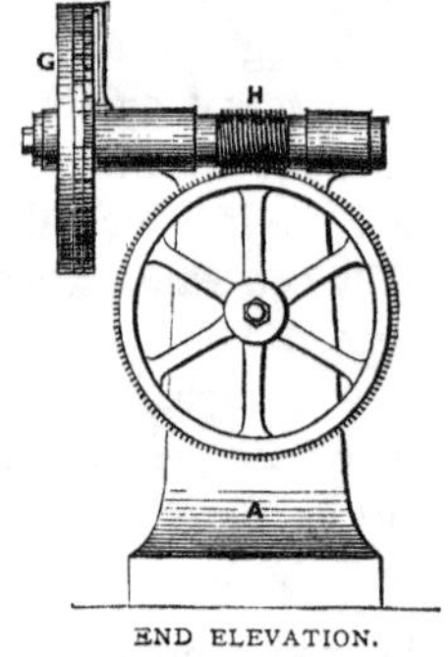

END ELEVATION.

SECTION.

FIG. 1.—Whitworth's Millionth Measuring Machine.

inch. The other screw, which likewise has 20 threads to the inch, is driven by a worm-wheel of 200 teeth, into which gears a tangent-screw, H, having fixed upon its stem the graduated wheel G. The circumference of this wheel being also divided into 250 parts, a movement through one division corresponds to a traverse of $\frac{1}{20} \times \frac{1}{200} \times \frac{1}{250}$ = one-millionth of an inch on the part of the bar C. Fixed pointers enable the exact distance through which either of the wheels F or G is moved, to be read off, so that we have thus the means of detecting this extremely minute difference in the length of any bars,—if, at least, we can fulfil the important condition of causing the micrometer screws to exert a perfectly equal pressure in every case. The arrangement by which this equality of pressure is secured is one of very great simplicity and beauty. Between one extremity of the bar under comparison and the sliding bar, a small steel plate with truly plane and parallel sides is introduced. This plate is called the 'feeler' or 'gravity piece,' and its ends (E E) are drawn out so as to rest upon two supports fixed upon the sides of the bed. When little or no pressure is exerted upon the bar D, the feeler, if one of its ends be momentarily raised from the support, falls back again by its own weight; when, on the other hand, the pressure is at all considerable, it is either incapable of being raised without violence, or when lifted, does not return; the friction, in fact, between its own plane surfaces and those of the bars between which it is placed forming a delicate measure of the pressure to which they are subjected. When this pressure is just sufficient to keep the feeler from falling by its own weight, without interfering with its perfectly free motion when touched, the correct adjustment has been given to the instrument.

Suppose now that a proposed duplicate is to be compared with a standard one-inch bar. The standard (D) and the feeler (E E) are first placed in the positions shown in the figure, contact between them and the sliding bars being nearly established by turning the wheel F; after which the final adjustment is given with the wheel G. As soon as the feeler, on its end being lifted, remains suspended instead of falling back upon its support, the adjustment is known to be complete, and the position of the wheel G is accurately noted. Since the new bar is to be an exact copy of the standard, the coarse adjustment wheel F is left untouched, the standard being released by moving the wheel G only, which is again adjusted when the duplicate of which the length is to be tested has been laid in the groove. If the position of the wheel then prove to be the same as before, it is evident that the length of the bars is identical; but if not, the exact difference between them is given in millionth parts of an inch by the number of divisions by which the second reading differs from the first; a movement through one of these divisions being sufficient to release the feeler or again to arrest its fall when the adjustment of G is correct. This degree of delicacy will thus be seen greatly to surpass that of the measurements which have been obtained by reading line measures with the aid of powerful microscopes. As an instance of the extreme sensitiveness of machines of this kind it may be mentioned that the one represented in our engraving is capable of detecting the expansion in a one-inch bar which is produced by merely touching it for an instant with the finger; and in the larger machine before alluded to—if due precautions be taken to protect it from dust, moisture, and currents of air—momentary contact of the finger-nail will suffice to produce a measurable amount of expansion in an iron bar 36 inches in length; a space corresponding to half a division on the fine-adjustment wheel, or one-two-millionth of an inch, having been rendered distinctly perceptible by it. Indeed, the expansion with every slight increase of temperature constitutes the only difficulty with which Sir Joseph Whitworth has now to contend, and he is of opinion that 62° Fahrenheit, which is adopted in this country as the standard temperature, is too low for the purpose, since, if it were increased to 70° or 80°, its uniformity would be much less liable to disturbance from the warmth of the operator's body. But it must not be supposed that this acme of mechanical precision is by any means easy of attainment. For it is obvious that if the slightest want of uniformity of pitch were to exist in the micrometer screws, or if the sides of the bars, &c., were not the closest possible approximations to true plane surfaces,—their ends being at the same time accurately at right angles to the common axis of the bars and screws—it would have been impossible even to approach the wonderful degree of accuracy which can now be obtained with these instruments.

For expressing the minute fractions of an inch which these measuring instruments enable us to take into account, the ordinary binary divisions (into eighths, sixteenths, &c.) are wholly unadapted. The use of a medley of such denominators as 64, 128, 256, &c., could only result in hopeless confusion and mistakes, so that in practice the inch is rarely divided on this system into more than 32 parts. That these are insufficient, even for ordinary work, is proved by the not unfrequent occurrence of such dimensions as '1-16th (full)' 'or 3-32nds (bare)'; and for accurate measurements, whether theoretical or practical, a decimal system is recognised as an absolute necessity. Unhappily, opinion is divided as to the best basis for such a system. Advocates of the mètre urge that, as a change is admitted to be necessary, we ought not to lose the opportunity for abolishing the whole of our present anomalous weights and measures, and substituting a system which is complete in itself, and which would greatly facilitate our transactions with France and other countries in which it is used. On the other hand, it is pointed out that by retaining one of our present units of length and merely subdividing it decimally, the difficulty of making the change, and by consequence the time required for it, is incomparably reduced; and that as far as the intrinsic value of the mètre as a theoretically perfect and recoverable unit is concerned, we should by no means be gainers by its adoption. It must be allowed that for the investigations of the chemist, and for similar purposes, a complete decimal system is invaluable, to which the fact that metric weights and measures are now almost exclusively used in such cases, bears testimony; but it by no means follows that they are equally well adapted to the requirements of the engineer or mechanician, nor that any real advantage whatever would be gained by compelling the one to conform to the practice of the other. Those who desire the overthrow of our present system for the sake of uniformity can hardly be aware of the sacrifices they would impose upon the manufacturing industry of the United Kingdom, nor can they consider the extent of the change, which for the time being would carry the inconvenience of a mixed system throughout the enormous area over which the English language is spoken. Moreover the loss which would be caused by the alteration or replacement of all the present gauges and measuring apparatus, and of many of the machines and tools used in our workshops, which would in itself be sufficiently considerable, would probably be a less serious matter than the overthrow of all the practical data and rules of calculation by the use of which we have arrived at our present superiority in mechanical manufactures; so that there is little doubt that the disadvantages involved in the adoption of the mètre, or any other new standard, would much more than counterbalance the advantage of having the same language of measure as our neighbours.

These and other considerations led Sir J. Whitworth to suggest the retention of the inch—which is in fact *the* unit of length for mechanical purposes—its subdivisions only being changed from vulgar to decimal fractions. This simple alteration provides us with the means of recording measurements of any required degree of minuteness, and so far from provoking the opposition which the attempt to implant an entirely new system would encounter, it has already made great progress towards general adoption.

Turning now from the consideration of the terms in which our measurements are expressed, to the practical means employed for their determination, the first thing which claims our attention is the Measuring-rule. Fig. 2 represents the form most commonly employed by mechanics, which consists of a strip of box or lance-wood 2 feet in length, divided throughout into inches, half-inches, quarters and eighths—occasionally into sixteenths—jointed at the centre, so as to enable it to be carried in the pocket. At the back of these rules there is frequently a brass slider, 12 inches long, which is marked with inches on its interior and with logarithmic scales on its exterior surface. The former are useful for enabling the workman to measure the depth of a groove or cavity; but he is seldom at the pains to make himself acquainted with the latter, though a few simple lessons in it would teach him that he carries in his pocket many hundred times as much power of calculation as he has in his head. The cheap rate at which these rules are sold is the only excuse for their usual inaccuracy. But various other forms of rule, made of wood, ivory, and metal, far too numerous for mention here, are to be found at every tool and instrument maker's shop; of which—as is the case with most other articles—the price generally regulates the quality. In order to ensure greater accuracy of measurement than can be expected from the cheap rules with which workmen usually provide themselves, and also to guard against the error which may result from the use of mixed dimensions (some quantities such as 1′ 1″ and 11″ being liable to be mistaken for each

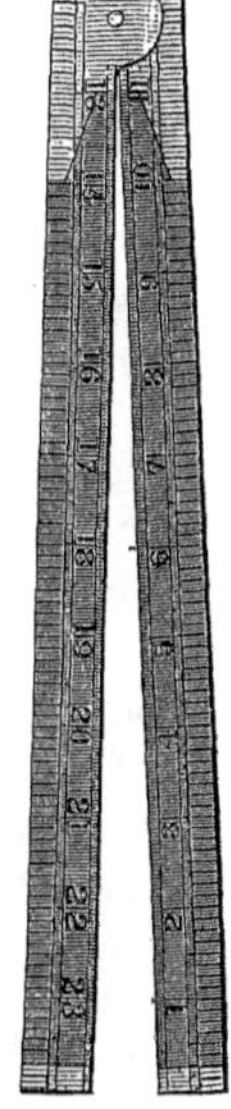

FIG. 2.
Two-foot Rule.

other), Sir Joseph Whitworth provides his workmen with well-made 20-inch rules, divided decimally throughout their whole length; all his working dimensions being expressed in inches and decimals only. The cost of these rules is a mere trifle, and the simplicity of their graduations renders them much more easy to use than those of the ordinary kind, which are frequently rendered complicated by having the inches on one edge divided into eighths, on another into tenths, on others again into twelfths and sixteenths, and in addition have the foot divided decimally.

Larger dimensions, when required but roughly, are most readily taken from a measuring tape. To obviate the tendency of linen tapes to stretch, steel ones have been introduced; but for accurately setting out long lengths, they are never likely to displace the simple expedient of using a pair of deal measuring rods each of the exact length of 5 or 10 feet, which are alternately placed end to end the requisite number of times. For still greater lengths as in measuring land, &c., the surveyor's chain either of 66 or 100 feet, is exclusively used—but this does not come within our present subject.

With straight and inflexible measuring scales or rules, however, it is evident that many of the dimensions which are required in the workshop, cannot be obtained directly. This is true of almost all cylindrical, as well as of many other forms of frequent occurrence, the most simple method of determining the size in such cases being to apply to the inaccessible portion of the work a pair of *callipers*, which ordinarily consist simply of two curved steel legs (Fig. 3), stiffly jointed together after the manner of a pair of compasses. The points of these are set to such a distance apart, that they are just able to pass freely over the cylinder or other article to be measured, its

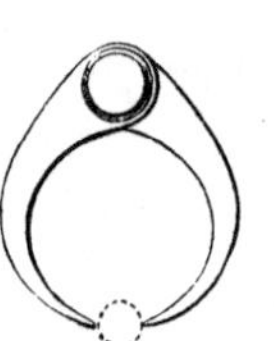

FIG. 3.—Callipers.

diameter or thickness being then found by laying the callipers upon an ordinary rule. It will be observed that the pair shown in the engraving can be used not only for outside dimensions, but also for inside ones, provided that the size of the aperture is sufficient to admit them. Two other forms of plain callipers will be found represented in Fig. 110; and yet others are made, amongst which may be mentioned *sliding callipers*, in which the pivoted joint is dispensed with, and a graduated sliding bar substituted. These, which somewhat resemble the thousandth gauge, Fig. 9, are true measuring instruments of considerable delicacy, and are not open to the charge of inadequacy for any but rough measurements. This, on the other hand, may justly be urged against callipers of the ordinary kind, which, although tolerably satisfactory correspondence between two pieces of work may be obtained with their assistance, are of comparatively little use for determining their true size. To do this entails the conversion of the end or contact measure, which is expressed by the distance between their points, into the line-measure marked on a scale or rule—a task by no means easy even in the case of rough measurements, and one which, when scientific accuracy is necessary, becomes a serious difficulty.

It may be mentioned in passing that the following is the only way of doing this correctly which has yet been devised. A line is drawn across each of two end-measuring bars, at or near their centres; their ends are then placed in contact with each other, and the distance between the lines is read with a pair of microscopes, such as are used for line-measure. Both bars are then turned end for end, and the distance between the lines is again read; the mean of these two readings gives the length of each bar. Since, as we shall hereafter see, all workshop measurements except those of an approximate character are obtained by some form of contact measure, the difficulty of this operation forms a strong argument in favour of the dissemination of end-measure copies of the standard, by which private measuring instruments may be verified when necessary, as has been strongly recommended by Sir Joseph Whitworth.

In determining the thickness of metal plates, sheets, and wires, the necessity for some apparatus in which small differences could be taken into account must have long ago made itself felt. To these not only is an ordinary rule generally inapplicable, its subdivisions, even if it could be applied, are altogether insufficient for the purpose. In consequence of this, it has become the universal custom to measure these (at least when they are of no great thickness) by means of a *wire* or *plate gauge*, which is in fact a simple kind of contact measure, formed by cutting a series of parallel-sided notches of varying widths, round the edge of a small plate of steel. In default of any means of ascertaining the true width of these notches in parts of an inch or of a foot, they have come to be distinguished by a purely arbitrary series of numbers. The *Birmingham wire gauge* (one form of which is shown—one quarter of its real size—in fig. 4, the notches being distinguished by the numerals from 1 to 26), may be taken as the type of these gauges; but unhappily it is by no means the only one. Local requirements and the insufficiency of the 'B. W. G.' series (for by these letters the numbers which belong to this gauge are distinguished) have brought various others into use; and since in their case, as in its own, the sizes are arranged and numbered on no definite principle whatever, and no sort of correspondence exists among them, great confusion and want of uniformity in these gauges prevails. From this the only chance of escape seems to lie in their total abolition, and the substitution of some entirely new gauge, which shall be sufficient for all ordinary purposes, and whose distinguishing numbers shall be arranged in a rational manner. Thanks to

FIG. 4.
Birmingham Wire Gauge (¼).

Sir J. Whitworth—to whom we so constantly have to express our obligations, and to whose valuable 'Papers'[1] we are much indebted—this is now in a fair way to be accomplished. He has carefully compared the principal gauges now or till lately in use, and he has drawn up a table (a portion of which is given below), in which not only is a definite series substituted for the erratic mode of progression which prevails in the Birmingham and other gauges, but the numbers possess this immense advantage, viz., that they represent the actual sizes expressed in thousandth parts of an inch. By this simple alteration several gaps are filled up, so that we are enabled to gauge sizes for which the old gauges had no numbers, and in addition to this we are relieved from the inconvenience of having the same size expressed by two or more different numbers, which was frequently the case before. As an instance of it we may mention that No. 24 of the new gauge represents the following numbers in five of the old gauges:—No. 23 of the Birmingham Wire Gauge; No. 10 of the Birmingham Plate Gauge; No. 72 of the Lancashire Gauge; No. 12 of the Music-wire Gauge; and No. 6 of the Needle-wire Gauge; the size in each case being ·024 of an inch.

Although gauges of this kind generally consist of notched steel plates, more or less resembling Fig. 4, they may be, and occasionally are, made upon a different principle. Some of these have but one very slightly tapered opening, instead of being provided with separate notches for each of the thicknesses to which they are to be applied; the number of any wire or sheet of metal being determined by observing the distance to which it can enter the opening. A *V-gauge* of this kind is, or until lately was, in use at the Royal Mint. When properly graduated, these gauges are capable of measuring the diameter of wire with tolerable correctness, but they are not well adapted for other than circular sections.

[1] *Miscellaneous Papers on Mechanical Subjects*: By Joseph Whitworth, F.R.S.: Longmans, 1858.

In the Birmingham and similar gauges, the notches are made by drilling a series of holes at some little distance

Table of the Principal Wire and Plate Gauges.

DECIMAL GAUGE (Thousandths of an inch.)	Corresponding Numbers of the Old			DECIMAL GAUGE (Thousandths of an inch.)	Corresponding Numbers of the Old		
	Birmingham Wire Gauge	Birmingham Plate Gauge	Lancashire Gauge		Birmingham Wire Gauge	Birmingham Plate Gauge	Lancashire Gauge
1	...			45	...	15*	56
2	...			50	18	16	55
3	...			55	...	17*	54
4	36	1		60	17*	18	52
5	35	2		65	16	19	51
6	...	...		70	15*	21*	49
7	34	...		75	...	22	47
8	33	3		80	...	24*	45
9	32	...		85	14*	...	43
10	31	4		90	...	...	42
11	...	...		95	13	25	41
12	30	5		100	...	26**	38
13	29	6	80	110	12	27**	34
14	28	...	79	120	11	28	31*
15	...	7	78	135	10	31*	29
16	27	8	77	150	9*	34*	23
17	...	...	...	165	8	36*	19
18	26	...	76	180	7	...	13
19	...	9	75	200	6**	...	5
20	25	...	...	220	5	...	2
22	24	...	74	240	4*	...	C
24	23	10	72	260	3	...	G
26	...	...	71	280	2**	...	K
28	22	11	70	300	1	...	N*
30	...	...	68	325	...	...	P*
32	21	...	66	350	...	...	S*
34	...	12	64	375	00**	...	V*
36	20	13	62	400	...	...	X**
38	...	...	61	425	000		
40	19	14	59	450	0000**		
				475			
				500			

NOTE—Sizes which differ from those in the first column by more than ·002 of an inch are marked thus ** ; those of which the difference exceeds ·001, thus *. All others either correspond exactly, or are within ·001 of an inch.

from the edges of the plate, with which they are then connected by saw-cuts of varying widths, the sides of the cuts being afterwards made smooth and parallel by driving into each notch a hard steel drift of the exact thickness represented by the number with which the notch is to be marked. The decimal wire gauges to which the above table refers are made by Mr. Stubs, of Warrington, from standard flat gauges and in general appearance greatly resemble the ordinary Birmingham gauge.

The range and the uses of this and the other gauges there mentioned may be briefly stated as follows:—

The first column gives the numbers of the Decimal gauge, which progress by regular steps, and in all cases represent the actual size, as already stated.

In column 2 is the entire series of the Birmingham Wire Gauge, from No. 0000 to No. 36, with one exception (No. 0). In the gauges for ordinary use (as in that represented in Fig. 4), the first four and the last ten sizes are generally omitted, the total number of notches being thereby reduced from 40 to 26. This gauge is much more extensively used than any of the other arbitrary gauges—wire (with some exceptions), sheet iron, and steel, and frequently also other sheet metals, being referred to it.

In column 3 will be found most of the numbers of the Birmingham Plate, or Metal gauge. Whenever the thickness of sheet metal—with the exception of those mentioned below—is expressed by an arbitrary number, this series is presumed to be referred to, although the numbers in the previous column, with the affix 'B.W.G,' are not unfrequently substituted for them. Even when the lower numbers of the Plate gauge are used, the series is often considered to end at No. 24, the larger sizes being then borrowed from the previous column. Since the numbers of these two gauges run in opposite directions, the whole arrangement is well adapted for producing confusion.

Column 4 contains many of the numbers and letters of the

Lancashire gauge. Commencing with No. 80, which corresponds with No. 29 B.W.G., this gauge absorbs all the numerals down to No. 1, following them up with the whole alphabet from A to Z; after which, still unsatisfied, it recommences with A.1., B.1., &c., till it reaches its maximum, H.1. (=·494, or nearly half-an-inch). It is used for pinion wire, and for round *bright* steel wire; 'Music wire' and 'Needle wire' being exceptions which have distinct and separate gauges of their own. 'Rope wire' and 'watch-spring wire' are also similarly favoured.

Among sheet metals the following are the principal ones which are not referred to the Birmingham Plate gauge:—

1st, Sheet iron and steel, as already mentioned.

2ndly, Zinc, which being chiefly rolled in Belgium, has a gauge of its own, which we give below, sheets of greater thickness being referred to the B.W.G. numbers, although these run in the reverse direction.

3rdly, Lead, which is estimated by the number of pounds per superficial foot, advancing generally by single pounds from 4 lbs. to 12 lbs.

4thly, Copper and Brass, which, when in large sheets measuring 4 feet by 2 feet (containing therefore 8 superficial feet) is described by the number of pounds in a sheet. This, however, does not apply to the 5-inch and other narrow widths of rolled brass, copper, &c., of which the thicknesses are expressed by the numbers of the B.W.G.)

5thly. Tin plate; which is not referred to any gauge, being sold in boxes, of which the number of sheets contained in each, their weight, size, and the marks by which they are distinguished, are given in the subjoined Table.

Lastly, Plates of iron or steel exceeding 1-8th (=·125) of an inch in thickness, are specified by the number of inches and binary divisions of an inch in the width of their cross section, being rolled (in ordinary cases) to eighths, sixteenths, &c. This enables the weight of any wrought-iron plate to be easily calculated by applying the useful rule which teaches us that—Rolled iron weighs 5 lbs. per superficial foot for every eighth of an inch in thickness.

Belgian Zinc Gauge.

No.	1	2	3	4	5	6	7	8	9	10	11	12	13
Equivalent in parts of an inch . .	·004	·006	·008	·009	·011	·013	·015	·017	·019	·022	·026	·030	·034
No.	14	15	16	17	18	19	20	21	22	23	24	25	26
Equivalent in parts of an inch . .	·037	·041	·045	·052	·059	·067	·074	·082	·089	·097	·104	·111	·120

Another example of contact-measuring instrument, and one which is to be met with in every workshop in which the accurate measurement of cylindrical work of moderate size is frequently required, is to be found in the case-hardened cylindrical gauges, of which an example is given in Fig. 5. To enable these gauges to be used for measuring both internal and external diameters, they are made in pairs, being called respectively *External* and *Internal Gauges.* Provided that they are correct in the first instance, no more ready or more certain method of testing cylindrical work can be conceived, — their simplicity of form and ample strength being for practical use a very high recommendation. But each gauge can of course only be used for the particular size to which it is made, so that a very large number of them is generally required. Indeed, a complete set of these gauges is an important item in the first outlay upon a fitting shop. Consequently an original set of standard gauges seldom is, and never should be, used, except for the production and correction of duplicates, the latter only being handed over for the use of workmen. But even when a full complement of these gauges has been provided (of which an ordinary set

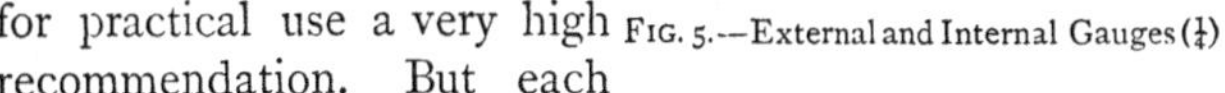

FIG. 5.—External and Internal Gauges (¼)

Sizes and Weights of Tin Plates.

Mark	Name	No. in each box	Size in inches	Weight of each box		
				cwt.	qrs.	lbs.
C I .	Common No. 1 . .	225	13¾ × 10	1	0	0
C II .	Common No. 2 . .	,,	13¼ × 9¾	0	3	21
C III .	Common No. 3 . .	,,	12¾ × 9½	0	3	16
X I .	Cross No. 1 . . .	,,	13¼ × 10	1	1	0
XX I .	Two Cross No. 1 . .	,,	,,	1	1	21
XXX I .	Three Cross No. 1 . .	,,	,,	1	2	14
XXXX I	Four Cross No. 1 . .	,,	,,	1	3	7
CD .	Common Doubles . .	100	16¾ × 12¼	0	2	21
XD .	Cross Doubles . . .	,,	,,	1	0	14
XXD .	Two Cross Doubles . .	,,	,,	1	1	7
XXXD .	Three Cross Doubles .	,,	,,	1	2	0
XXXXD	Four Cross Doubles . .	,,	,,	1	2	21
CSD .	Common Small Doubles .	200	15 × 11	1	2	0
XSD .	Cross Small Doubles .	,,	,,	1	2	21
XXSD .	Two Cross Small Doubles	,,	,,	1	3	14
XXXSD	Three Cross Small Doubles	,,	,,	2	0	7
XXXXSD	Four Cross Small Doubles	,,	,,	2	1	0
WC 1 .	Wasters, Common No. 1 .	225	13¾ × 10	1	0	0
WX 1 .	Wasters, Cross No. 1 .	,,	,,	1	1	0

rises by either 10ths or 16ths for the first inch, by 8ths for the second, thence by quarters up to 6 inches,—beyond which they are not usually made) uneven dimensions will be frequently found to occur, for which they render no assistance. In these cases a measuring machine, such as that shown in Fig. 10, will be found to be a most valuable adjunct, as it enables the number of standard gauges to be greatly reduced, whilst at the same time any intermediate dimensions can be easily and accurately obtained. By grouping together the successive numbers of a series of external gauges, so that about 8 of them are formed in one piece, their cost is considerably reduced. Thus modified, they are known as *Stepped Gauges*, one of which is represented in Fig. 6; but their construction renders them

FIG. 6.—Stepped Gauge (¼).

of more limited application than the gauges last described, and the smaller amount of cylindrical surface devoted to each dimension renders them liable to more speedy deterioration. Moreover, as they are not usually provided with collars, they cannot be used for outside measurements.

The cylindrical gauges would be too delicate if made smaller than one-tenth of an inch diameter, so that for sizes below this down to one-fiftieth of an inch *Flat Gauges* are

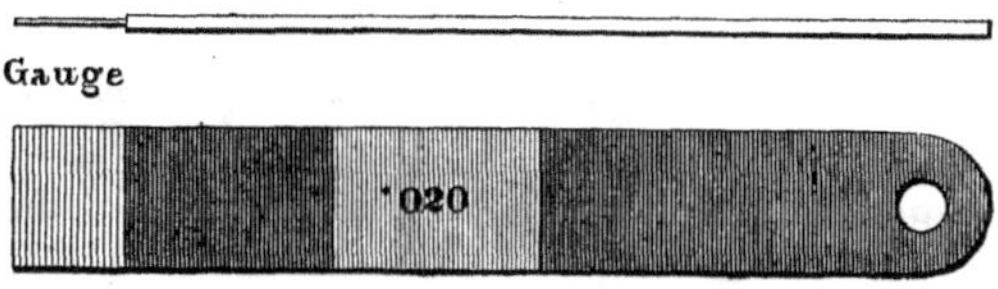

FIG. 7.—Flat Gauge (full size).

made, one of which is shown in Fig. 7; the two faces of the gauge are made true parallel planes. A separate gauge of this form was made by Sir Joseph Whitworth for each notch in the Decimal Wire gauge before referred to.

Before, however, proceeding to notice the measuring machine just alluded to, it will be well to call attention to one or two delicate, but very portable instruments, which, although they are not often to be met with in a workshop at present, might be more extensively used there with great advantage. To a careful workman they are capable of rendering valuable assistance.

The simplest mechanical means by which minute quantities can be rendered visible would probably be an arrangement by which the forward motion of a sliding wedge should be made to record every slight increase in its thickness, and instruments on this principle have been proposed by Mr. Ramsbottom. Their magnifying power, as it were, depends upon the amount of 'taper' given to the wedge; for instance, a wedge 10 inches long and of 1 inch total thickness would have a sliding motion of 1-10th of an inch for each 1-100th of an inch by which its thickness varied. A modification of them has been used in the locomotive engine shops of one of our large railway companies; the wedge being bent into a circle of about 10 inches circumference. Each complete revolution corresponding to a variation in its thickness of 1-4th of an inch, 1-1000th of an inch can be measured by causing it to revolve through about ·04, or 1-25th of an inch, which is a distinctly visible quantity. In most cases, however, this instrument cannot be directly applied to the work, and measurements must be made through the intervention of callipers. Altogether this arrangement of the screw—for such the wedge becomes when thus modified—is far inferior in point of convenience to that next described.

Fig. 8 is a half-size drawing of a pocket instrument made by Messrs. Elliott Brothers, which is capable of measuring very minute quantities. As ordinarily made the screws have 50 threads to an inch, their heads having 20 divisions on the circumference. Each division therefore corresponds to a longitudinal motion of 1-1000th of an inch on the part of the screw; but this by no means represents the limit of delicacy which can be obtained on this system. It should be noticed that any slight error in the number of threads per inch which there may be in the screw, can be corrected in graduating one of these instruments by recording the number of its revolutions on a spiral line, instead of on one parallel to its axis. The milled portion of the head is sometimes provided with an independent 'Breguet' motion, which obviates any tendency to put too much or too little pressure on the article which is being measured.

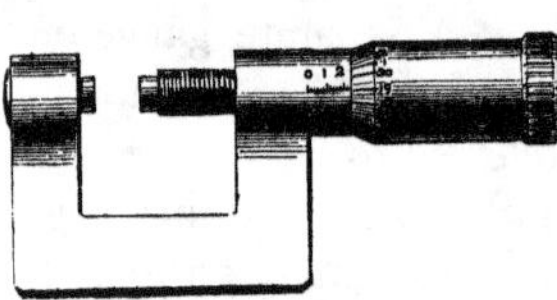

FIG. 8.—Thousandth Gauge (½).

Another form of instrument, which may be carried in the pocket, is shown—also half size—in Fig. 9, the original from which our engraving has been taken having been made by Mr. Holtzapffel. The scale A is marked with inches, which are divided into fiftieths. By means of the vernier B each fiftieth can be divided into 20 parts, the distance between the inside knife-edges of the jaws C C being thus read off to the thousandth part of an inch. These jaws being each

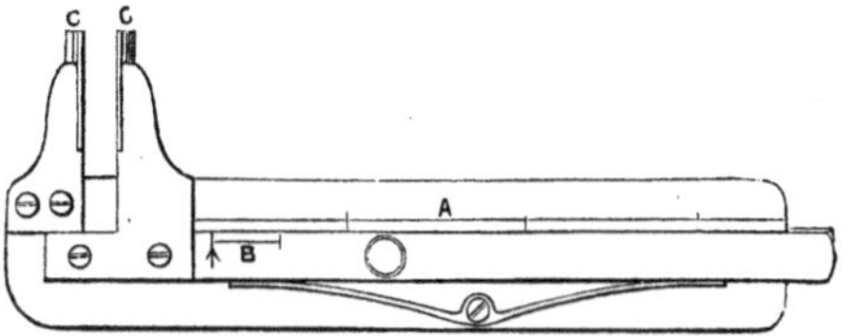

FIG. 9.—Sliding Thousandth Gauge (½).

made exactly 1-10th of an inch wide for a portion of their length, they can be used to some extent for inside dimensions, two-tenths of an inch being in this case added to the reading given by the scale. Either this or the previous instrument can be used for all the purposes to which the complicated wire and plate gauges above mentioned are applied, and for many others in addition.

But for verifying working copies of gauges—for ensuring the accuracy of the various templates or patterns of size which are in daily use in the workshop—for maintaining constant uniformity of measurement, so that perfect agreement shall obtain between any articles of the same dimensions, even though produced at distant periods of time; for all these, and for many similar purposes, the measuring machine, with a brief notice of which we will conclude this chapter, leaves nothing to be desired. It has been specially designed by Sir Joseph Whitworth for workshop use, and, as the great importance of accurate measurements in machine work becomes more and more generally appreciated, it will doubtless be found to be an increasing necessity therein.

The reader who has followed us through the description

of the millionth measuring machine (Fig. 1,) given at a previous page, will be able without assistance to understand the general construction and method of using the less delicate instrument of which Fig. 10 is a perspective view. We shall perhaps therefore best describe it by calling attention to the chief points in which it differs from the previous machine. In the first place let us call to mind the difference in the duties which each of them has to perform. With regard to delicacy of measurement their respective requirements are as far removed from one another as is the accuracy demanded

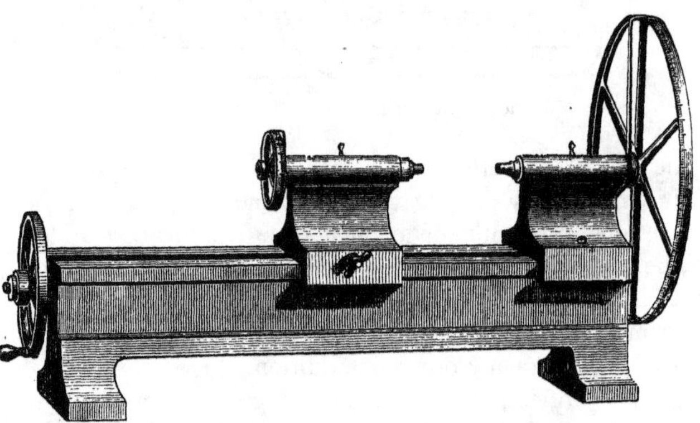

FIG. 10.—Machine for Measuring to the Ten-Thousandth part of an inch.

of a scientific standard from that which suffices for commercial purposes; so that whereas in the one case it is of the greatest importance to be able to record the most minute quantities—we have seen how the millionth part of an inch can be rendered visible, and if it were possible it would doubtless be desirable to take into account yet smaller fractions—in the other, it is useless to provide the means of measuring distances, of which it is at present impossible for us to take cognisance in mechanical handicraft. On this account this latter machine is provided with graduations, each of which corresponds only to one ten-thousandth part of an inch. But the principal differences in its appearance are due, not to this diminution in magnifying power, but to the necessity for enabling this machine to admit articles of greatly varying size and form; since, if it could be applied only to measuring the length of bars of one particular section, it would be of comparatively little use in the workshop. On this account the headstocks, which in Fig. 1 are cast in one piece with the bed of the machine, are here separate castings; one of them being fixed at the end of the bed, the other capable of being moved to any part of it, and clamped there. The bed bears much resemblance to that of a small turning lathe, as also does each of the headstocks to the poppet, or sliding headstock, by which the back centre is supported. This movement of the headstock, which is effected by turning the small wheel at the left-hand side of the figure, rotation being thus imparted to a quick-threaded screw within the bed, enables the ends of the sliding bars to be placed at one foot or at any smaller distance from one another, whilst—owing to the height at which they are supported above the bed—a cylinder of as much as six inches in diameter can be brought into position for measurement between them.

The graduated portions of the instrument are three in number—in the first place the inches from 1 to 12 are marked upon the surface of the bed, so that the moving headstock can at once be approximately placed at the required distance; secondly, the small wheel carried by this headstock has 250 divisions round its circumference, each complete revolution of this wheel corresponding to a backward or forward movement of one-twentieth of an inch on the part of its sliding bar (that being the exact pitch of the screw to which it is attached); and lastly, the large wheel of the fixed headstock has 500 graduations upon its rim. The screw to which it is attached having also 20 threads to the inch, rotation through the space of one of these graduations causes the sliding bar in this case to move through $\frac{1}{20} \times \frac{1}{500}$ = one ten-thousandth part of an inch. The capabilities of this machine will thus be seen to be far more extensive than those of the other and the more sensitive instrument, and although its performance may be less calculated to excite surprise and admiration, it is far better adapted to the fulfilment of every-day requirements. In using it standard bars and cylindrical gauges are of course required, this, like the former, being intended for making comparisons rather than original measurements; but, as we have before pointed out, a small number of carefully kept standards only are necessary, since intermediate steps in the series can be readily filled up, owing to the facilities which the graduated wheels afford for increasing or diminishing, within considerable limits, any dimension which has been accurately obtained from a standard bar or gauge. In the comparison of bars, a simple form of support for them between the headstocks is desirable, since a 'feeler' can then be used in the manner already pointed out; but in the case of cylindrical gauges, or portions of work, the operation can be easily effected without such assistance, the gauge or other article being merely held in the hand and passed between the ends of the sliding bars. When the true adjustment has been given to the machine it has been found that an alteration of one forty-thousandth of an inch in the distance between the bars causes a distinctly perceptible increase or decrease in the resistance which the cylinder encounters in passing between them. Large gauges, &c., which are too heavy to be conveniently held in the hand, may be suspended vertically over the machine: the adjustment can then be given as correctly as in the case of the very smallest articles.

A frequently recurring example of the advantage of fine measurements in mechanical work, which will serve well to illustrate the use of this machine, occurs in fitting together two portions of a piece of work—such as a wheel upon an axle. If it be required to revolve, the diameter of the bearing must be somewhat larger than that of the axle; if to be driven on so as to fit tightly (as unhappily is the case in railway wheels), the diameter must be somewhat smaller. For either purpose the best result only is obtained with one particular amount of difference between the external and internal diameters; and although this can only be learnt by experience, yet, when learnt, it should for the future be adhered to. Let us take the case of an external cylindrical gauge being required for this purpose 4·003 inches in diameter —that is to say, one which shall exceed that of a standard 4-inch gauge by three-thousandths of an inch (and it may be mentioned that one-thousandth more or less makes all the difference between a good fit and a bad one). The moving headstock of the machine having been clamped at the 4-inch division on the bed, its wheel is adjusted till the 4-inch standard gauge can just pass freely between the ends of the sliding bars; the largest wheel upon the fixed headstock having been previously set to 0, or zero. This wheel is then moved through 30 divisions, which gives the exact difference required in the distance between the sliding bars. Perfect but free contact with their ends on the part of the proposed 'difference gauge' then proves its correctness.

Difference Gauges are usually made in sets of three, those used for gauging the bores of rifles, rising by one 5000th of an inch.

CHAPTER II.

ON HAND-TOOLS USED FOR WOOD.

IN his well-known work on 'Turning,'[1] &c., Mr. Holtzapffel divides the cutting-tools used by hand into three classes, viz., paring tools, scraping tools, and shearing tools. He admits, however, that this classification is open to criticism, and although we shall be much indebted to Mr. Holtzapffel in the course of this and some of the succeeding chapters, and are glad to bear testimony to the value of his work, we may be pardoned for noting at the outset one or two objections to this mode of grouping.

Paring, or splitting tools, are therein defined as having 'thin edges, the angles of which do not exceed sixty degrees; one plane of the edge being nearly coincident with the plane of the work produced (or with the tangent, in circular work).' 'Scraping-tools,' on the other hand, have 'thick edges, that measure from sixty to one hundred and twenty degrees. The planes of the edges form nearly equal angles with the surface produced; or else the one plane is nearly or quite perpendicular to the face of the work.' Yet joiners' planes are placed in the first group, of which the small facet produced in sharpening the iron on an oilstone is said to form an angle of 10° with the surface of the work; and saws are placed in the second, although the back of each tooth may also be inclined 10°, or even only 5°, to the cut which it produces. Moreover, under the above definitions, even a wood-chisel, the type of paring tools, loses its title to being classed among them, when, in the hands of the

[1] *Turning and Mechanical Manipulation.* By Charles Holtzapffel: London, 1843

turner, it is applied perpendicularly to the surface of the work, as is sometimes the case. Thick edges, as distinguished from thin ones, seem to offer but little help out of these difficulties of classification. Under the head of 'paring-tools,' we find 'most of the engineer's cutting, turning, and planing tools for metal,' in spite of the thickness of their cutting edges, whereas razors (although Mr. Holtzapffel consistently includes them among paring tools) ought surely, if they deserve their name, to be classed among scrapers.

A better distinction would probably be one in which, in addition to the inclination of the tool, the direction of the force applied to it was taken into account; though whether on this, or on any other basis, a rigid classification could be carried out, seems very doubtful. Certain it is, that at least in cases where continuous rapid motion is imparted to a tool, or to the work with which it is brought into contact—as, for instance, in the case of a circular saw, the cutter of a rotary planing-machine, wood in the process of being turned, &c.—the effects produced are quite as dependent upon the speed, as upon the angle and inclination of the cutting edge; a fact which must not be lost sight of in reducing to definite groups the miscellaneous collection of implements comprised under the head of 'cutting-tools.' No such attempt, however, will be made in this and the following chapter, in which it will be impossible for us to do more than to glance briefly at the more ordinary types of hand-tools—some insight into their principles being essential to the proper understanding of the action of the various machine-tools which form the main portion of our subject.

It should be observed that simplicity in the form of a tool by no means implies facility in its use. Indeed, the very reverse of this is much nearer the truth, and within certain limits it may be said that the more simple the tool, the greater is the skill required to use it properly. An extreme instance of this will be found in the *scriber*, which is so simple in its form as doubtfully to rank among cutting-tools. It consists merely of a straight pointed piece of steel of circular section, and it is used by workmen for marking on metal, just as a pencil is used for marking on paper. The working part of it being merely a point, it has, when used alone, no tendency to move in one direction more than in another: the 'guide principle,' as it has been termed, being entirely absent. A straight, or curved line, can therefore only be described with it, if the workman is sufficiently skilful to guide it perfectly correctly with his hand, which in mechanical work is out of the question. A scriber is in consequence invariably used in conjunction with a guiding edge of the required form—the difficulty of using it alone being for all practical purposes insuperable.

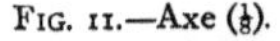

Fig. 11.—Axe (1/8).

Neglecting metal tools for the present, we will confine our attention to those used for wood. Foremost among them is the *axe* (Fig. 11), the general form of which seems to have undergone but little change since the date of flint implements, to some of which the axe-head—more especially the American pattern (Fig. 12)—bears no inconsiderable resemblance. This pattern, however, which is less cumbrous than the ordinary English axe, has come considerably into use in this country. Both of these are sharpened by being ground equally on each side of the cutting edge.

Fig. 12.—American Axe-head (1/8).

When it is necessary to produce a comparatively flat surface with an axe—as, for instance, in squaring logs of timber—a modified form, known as the *side-axe*, is used, of which the edge is ground on one side only. The flat, or unground side, then forms a guiding surface, which enables successive cuts to be made with an amount of precision unattainable with an ordinary axe.

Still greater accuracy can be obtained with the *adze* (Fig. 13), though this is in great measure due to the manner in which it is held. Standing upon his work, and planting his blows with wonderful effect upon the timber beneath his feet, a good workman wields his adze with such certainty, that he can split the sole of his slipper without fear of even grazing the skin of his foot. An equally good tool for small work is the *hand-adze* (Fig. 14). In this the handle is attached by an iron strap, in a manner quite as effective as the usual plan of driving it upwards through an eye, and it can be liberated by a single blow from a hammer. Although made in Sheffield for exportation, this tool has never come into use in this country, though it might do so with advantage. Indeed, in the use of most of the varieties of axe, we must confess inferiority to our trans-Atlantic and other neighbours, who export instead of importing their timber.

FIG. 13.—Adze ($\frac{1}{8}$).

FIG. 14.—Hand-Adze ($\frac{1}{6}$).

For the magnitude of their effects in comparison to the amount of energy expended, the various forms of axe and adze certainly stand unrivalled amongst wood-tools, combining as they do the cleaving power of a wedge, with the force of impact of a hammer. They are, therefore, universally applied for heavy timber work, in the preliminary stages of which they give invaluable assistance. Their edges, however—more especially those of ordinary axes—have so small an amount of guiding surface, that a very high degree of skill on the part of the operator is necessary, in order to make the cut in the desired position and direction. It is of course advantageous for the edges of these tools to have as small an amount of thickness as is consistent with sufficient strength to resist the treatment to which they are subjected; the angles of their cross sections therefore vary from about 25° to 40°, according to circumstances, those of the side-axe and adze generally approaching the smaller, and that of the common axe the larger of these angles.

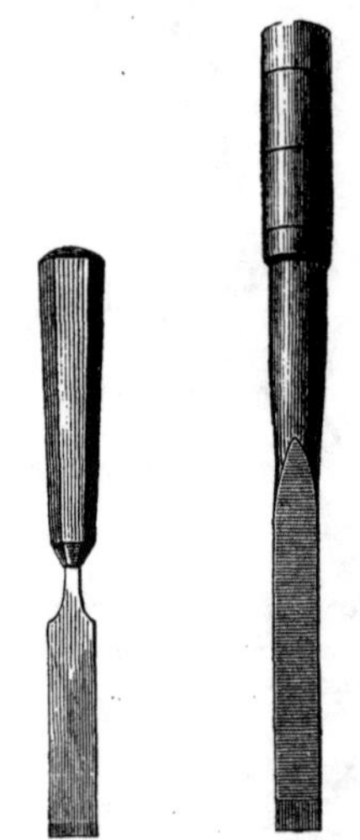

FIG. 15. Firmer Chisel (1/8). FIG. 16. Socket Chisel (1/8).

Figs. 15 and 16 show respectively a cast-steel *firmer chisel*, and a steel-faced *socket chisel*, which differ from one another chiefly in the provision for the attachment of their handles. Like the side-axe and the adze, chisels of this kind have 'single bevelled' edges, being ground on one side only. By keeping the unground side pressed closely against the work, the tendency of the edges to run into the material can be completely resisted; the even pressure with which a chisel is driven where thus used for paring, being especially favourable to this application of the 'guide principle.' In this manner small surfaces can be made almost as flat as with a plane; but it should be observed that this can only be done as long as the original flatness of the unground face of the chisel is not interfered with.

In the *mortice chisel* (Fig. 17), which may be either firmer or socket-handled, the guiding power of the unground face is also taken advantage of in cutting the vertical sides of mortices. Chisels of this kind have considerable extra thickness given to the blade, to enable them to stand the rough usage which they are liable to meet with. The width of their edges and that of carpenters' chisels generally is rarely less than $\frac{1}{8}$th of an inch, or more than 2 inches. All of these are sharpened by grinding them on one side only, as already stated (but which cannot be too strongly impressed upon a beginner), at an angle of from 20° to 30°, according to the hardness of the material to be operated upon, a small facet at a somewhat greater angle (ordinarily exceeding the former by about 10°) being afterwards formed upon an oilstone, for the purpose of giving a smooth and clean cutting edge.

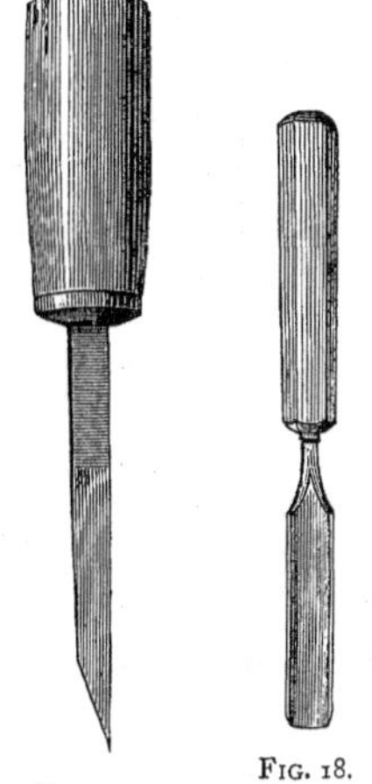

FIG. 17. Mortice Chisel ($\frac{1}{6}$).

FIG. 18. Carpenter's Gouge ($\frac{1}{6}$).

The *carpenter's gouge* (Fig. 18) is also either socket or firmer-handled, and it is ground and set in the same manner, and at much the same angles as the foregoing, but the bevel being generally on the convex side, its cut is much less easily guided.

Except for the removal of moderately thin or narrow shavings, the pressure of one or both hands is insufficient for working a gouge or chisel. It is then propelled by successive blows upon the end of its handle, for which purpose a *mallet* (Fig. 19) is much to be preferred to a hammer. The latter, whilst more destructive to the handle of the tool, is less effective in driving it into the work.

From the carpenter's paring chisel, an easy step leads us to the *draw-knife* (Fig. 20), for although they differ greatly

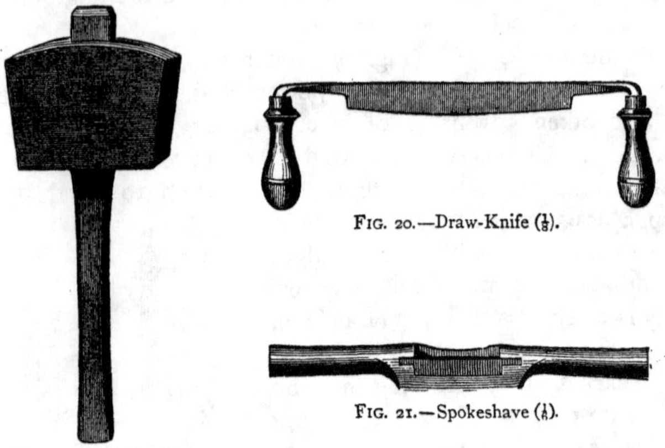

FIG. 20.—Draw-Knife (1/8).

FIG. 21.—Spokeshave (1/4).

FIG. 19.—Mallet (1/4)

in form, the depth of their cut is controlled in a precisely similar manner. This consists, as previously mentioned, in keeping the flat side of the blade towards the surface of the work, its tendency to run into it being either wholly or partially resisted by raising or depressing it. From the position of the handles, it is obvious that a draw-knife could not be applied to a surface whose width exceeded the length of the cutter; but, in fact, its use is much more restricted by the amount of power which is necessary to remove, even from soft wood, a shaving which is wide as well as thick. Neither this, nor any of the preceding tools, the depth of whose cut is dependent upon the inclination of the cutter, are competent to remove very thin, or uniform shavings.

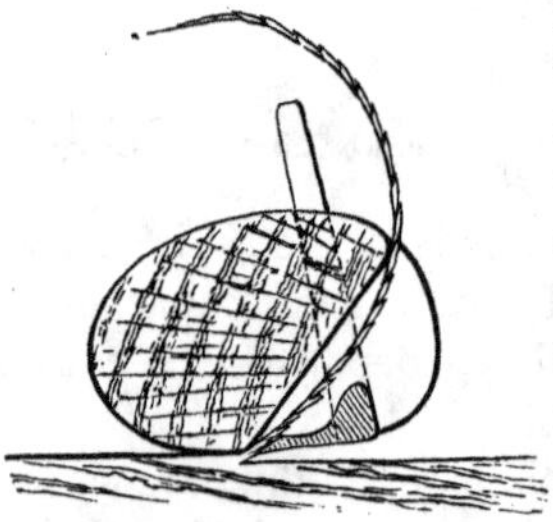

FIG. 22.—Section of Spokeshave.

With the *spokeshave* (Fig. 21) this is to some extent possible, for this tool will be found on examination to differ from the preceding in principle, much more than in form. Fig. 22 represents a section through the centre of a spokeshave, from which it will be seen that the piece of (beech or other) wood which carries the cutter, is not in this case merely a convenient form of handle; it is an integral part of the tool itself. Running along the whole length of the blade, it forms a guide which renders it impossible for the cutting edge to penetrate the work beyond a certain fixed depth. By regulating this depth we ought to be able to

FIG. 23.—Jack Plane (⅛).

remove with this instrument perfectly uniform shavings of any required degree of thinness. But in practice, a spokeshave is by no means as efficient a tool as it would appear to be, and it is never employed where much accuracy is required. The effect of use is to enlarge the *mouth* (*i.e.* the space between the wooden guide and the cutting edge), more especially towards the centre, where it gets most work. Grinding the cutter further increases this evil, and the cutter itself—of which the upturned ends are rather obstacles to sharpening—is liable to have its edge much thickened by the process. This, however, is partly prevented by forming a groove in its upper surface, which can be seen in the section.

In the various kinds of surface plane, the whole of these defects are remedied, and the result is a beautiful instrument, which for accuracy of performance surpasses all other hand tools. The general forms of the *jack plane* (Fig. 23), and of the *smoothing plane* (Fig. 24), are too well known to need description, but the requirements which have given rise to these forms deserve a little consideration.

FIG. 24.
Smoothing Plane (⅛).

In the first place, with reference to the *stock*, which is generally made of beech wood, and which carries the cutter, or *plane-iron*, as it is called. Through this stock there is a vertical aperture, of which the lower portion serves the same purpose as the wooden guide in the case of the spokeshave, and forms, with the cutting edge, the *mouth* of the plane. This, as we have seen, would be sufficient to regulate the penetration of the cutter, but it would not prevent it from following all the inequalities of the surface to which it is applied. Now the object of planing a surface is to render it flat, by removing its inequalities, and for this it is necessary that the cutter should act upon the projecting portions only. This end is simply and effectually attained by giving considerable length to the stock, which causes a plane, in operating upon a rough piece of wood, to remove successive shavings from the more prominent parts, until they are all brought down to the level of the deepest depression in its surface. When this has been done, a continuous shaving can be taken from the whole length of the piece, and the surface (if narrow) is approximately flat. How nearly it approximates to perfect flatness depends upon the skill of the workman, who knows, among other things, that he must try to plane it slightly 'hollow,' rather than 'round,' since, if he uses a plane sufficiently long in the stock, it is impossible for him to give any great amount of concavity to a surface of moderate size. For this reason jack planes, with which the first roughing-out of a surface is done, have as much length given to them as is possible, without making them cumbrous—*i.e.*, from about 14 to 17 inches. *Trying* (or true-ing) *planes*, whose office is to correct the inequalities left by the former, are from 22 to 24 inches, or—if intended for making long joints, in which case they are called *jointers* —28 to 30 inches; whereas smoothing planes, which have merely to give the finishing cut to the already flattened surface, are only about 8 inches long.

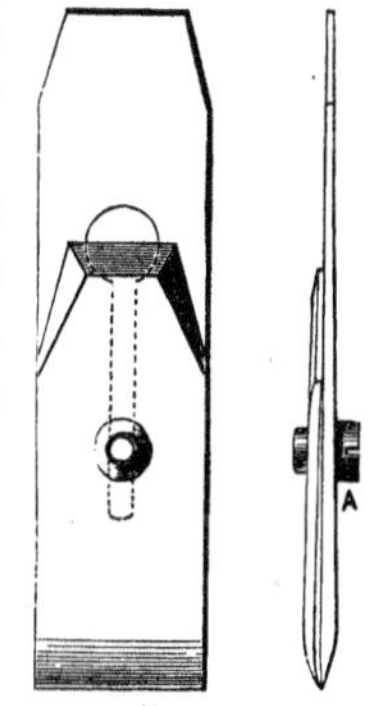

FIG. 25.
Double Plane Iron (¼).

The *plane-iron*, of which the lower part of one face only is made of steel, was formerly a simple chisel-like blade from ⅝ths to ⅞ths of an inch narrower than the sole of the plane; but this has been greatly improved of late years by attaching to its upper surface a second inverted blade of the same width as the first. Together they form the *double-iron* shown in Fig. 25, which for surface planes has almost entirely superseded the old single-iron. The *top-iron*, though it takes no part in the cutting of a shaving (and consequently when once sharp it never again requires sharpening), enables the bottom-iron to make a

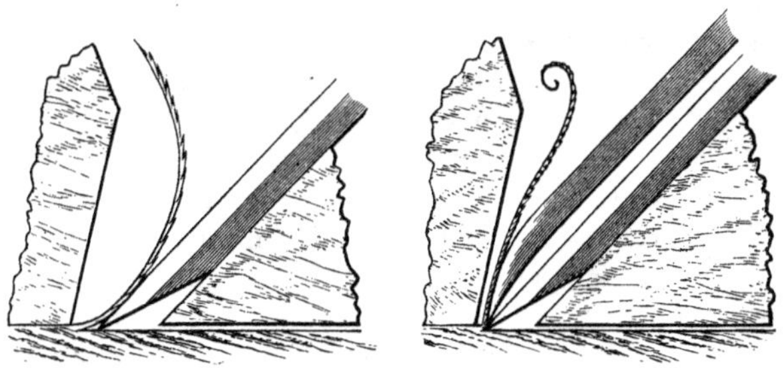

FIG. 26.—Sections of Mouths of Single and Double-Iron Planes.

much cleaner cut, more especially when the grain of the wood is unfavourable. Fig. 26, which shows in section the mouths of a single, and of a double-iron plane will explain the reason of this. It must be borne in mind that the tendency of a cut to run with the grain of the wood in advance of the cutting edge, which is sufficiently obvious in removing a thick shaving, exists also, though to a less degree, in the case of a thin one. This results, when the direction of the grain is unfavourable, in the continual tearing up of the fibres of the wood in front of the plane iron, to an extent which depends upon the thickness of the shaving and the frequency with which it is broken; for the more often this is done the less leverage can it exert in tearing up the succeeding fibres. Now an ordinary plane has a *pitch* of about 45°, *i.e.* the *bed* upon which the iron rests is at that inclination, and this angle is insufficient to break a soft-wood shaving if moderately thin; so that although the shaving is smooth, the surface of the work is liable to be left rough. As a remedy it might be suggested to increase the pitch; but this, though it would certainly improve the breaking-power of the iron, would tend to make it scrape instead of cutting. So recourse is had to the double-iron, in which the duty of the top-iron is simply to break the shaving as soon as possible after it has been cut. By means of the screw (A, Fig. 25) which keeps them in contact, it can be set at any required distance from the edge of the bottom-iron, but the more closely it is set the more often is the shaving broken, and the more power has the workman to expend in the operation. It is therefore set very close to the edge for finishing cuts, when the shavings are very thin, and farther from it when they are thicker, the maximum being about $\frac{1}{16}$th of an inch.

We mentioned above that the ordinary angle of the bed upon which the plane-iron reposes, is about 45° from the sole. This is known as *common-pitch*, and is adopted for all surface or *bench-planes* for soft wood. But for the harder kinds of timber, in the working of which a nearer approach to a scraping action is admissible, an increased pitch is adopted with advantage. Mr. Holtzapffel states that the angle of that known as *York-pitch*, which is used for bench-planes for mahogany, wainscot oak, and other hard and stringy wood, is 50°, that of *middle-pitch* 55°, and that of *half-pitch* 60°, these two being used for moulding planes for soft and hard wood respectively. The angle of the bevel produced by grinding the bottom iron of a plane is about 25°, the small facet formed upon the oilstone having—as in the case of a chisel—a further inclination of some 10°, so that the actual angle of the cutting edge is about 35°. Upon the maintenance of this angle, and upon the perfection of its edge, the performance of a plane-iron so greatly depends, that we shall do well to mention here the principal points which must be attended to in the important and frequently recurring operation of sharpening a bench-plane.

The requirements for this purpose are: first, a grindstone, flat as to its edge or 'face,' true, and of good quality; secondly, a good and flat oilstone. Of the various kinds of stone we shall have more to say hereafter. The wedge which holds the plane-iron having been released,—in the case of a smoothing plane by one or two blows from a hammer or mallet on the hinder end of the stock, in that of a jack or other long plane, by similar blows upon its upper surface near the opposite extremity,—the double-iron can be removed from the aperture in the stock. The upper iron should then be separated from the lower by loosening the screw (A, Fig. 25) and passing its head through the enlarged end of the slot in the lower iron. As already stated, the top-iron, when once brought to its proper shape, requires no further treatment whatsoever, its sharp edge being only necessary to ensure its perfect contact with the face of the bottom-iron, by which, when in use, it is completely protected. Laying this aside therefore, the operator confines his attention to the bottom-iron, grinding it when necessary with a single bevelled edge, inclined to the face at 25° or thereabouts, and in all cases finishing it by rubbing it upon the oilstone, at an inclination of about 35°. As in the case of a chisel, the face or unground side should either be left altogether untouched, or should at most be laid perfectly flat upon the oilstone, and be slightly rubbed upon it. No attempt should ever be made to grind this side of the iron or in any way to impair its flatness.

When an edge, keen, uniform, and at right angles to the sides of the iron, has been thus produced, the top-iron must be replaced, screwed up securely, and the double-iron be returned to the stock. The wedge which holds it being driven up lightly in the first instance, the plane is then 'set' by a succession of taps with a light hammer upon the end of the iron, or upon the stock, according as the projection of the edge below the sole is insufficient or excessive. By running the eye along the sole from front to back, the amount of this projection can be easily seen. It must of course vary with the thickness of the shaving which it is intended to remove, the wedge being driven up when the desired set has been obtained.

In course of time the mouth of a plane gets considerably enlarged, by which its power of breaking the shavings is diminished and its action interfered with. To remedy this, the part of the sole immediately in front of the cutter is occasionally made of iron. Cabinet-makers, who carry the use of the plane to the greatest perfection, use stocks which are wholly or in great part made of iron or brass. Some of these have a single iron inverted, instead of a double one, more especially if (like *mitre-planes*) they are intended for planing across the grain of the wood, when no special precautions for breaking the shavings are required. The mouth of a plane of this kind is shown in section in Fig. 27, the stock being entirely of brass with the exception of a piece

FIG. 27.
Section of Mouth of Brass Plane.

of wood upon which the iron is bedded. The angle of the bed is in this case only 21°, but the reversed position of the iron, which is ground at an angle of 20° and set at about 30°, makes it equivalent to a pitch of about 50°. The extreme fineness of the mouth coupled with this increase of pitch compensates for the absence of a break-iron in planing hard and cross-grained wood, even in the direction of its fibres, with a tool of this kind. In this as well as in the preceding sections the portion of the blade which is left white approximately indicates the portion of the 'iron' which is made of steel.

The various forms of plane for special purposes are too numerous even for mention here. Amongst them are: *moulding-*

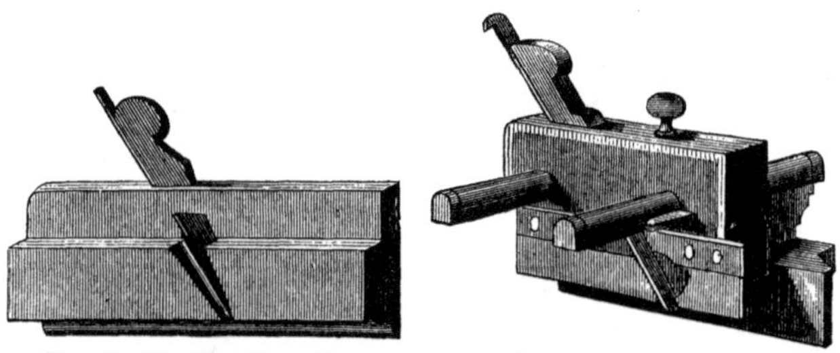

FIG. 28.—Moulding-Plane ($\frac{1}{8}$). FIG. 29.—Plough ($\frac{1}{8}$).

planes (Fig. 28), for 'running' various kinds of beads, flutings, ogee and other mouldings, which have their soles made of hard wood and formed to the converse of the mouldings they produce; *rebate-planes*, in which the lower part of the plane-iron is increased to the full width of the sole, and is frequently set obliquely; and many others, of which the irons are single, and the performance more or less defective. But since the introduction of wood-moulding machinery, their occupation has in great measure gone. A very useful but complicated form of plane is the *plough* shown in Fig. 29, which is capable of cutting grooves of varying widths, depths, and distances from the edge of the work.

But we must pass on to what is perhaps the most indispensable of all tools to the wood worker, viz., the *saw*. Though the forms of the teeth of different saws vary considerably, those in general use in this country may be referred to the three types shown in Fig. 30. The first of these (A) is the one usually adopted for the teeth of hand-saws, tenon- and dovetail-saws, and most of the others which are used single-handed, and it will best serve us in considering the action of saws in general.

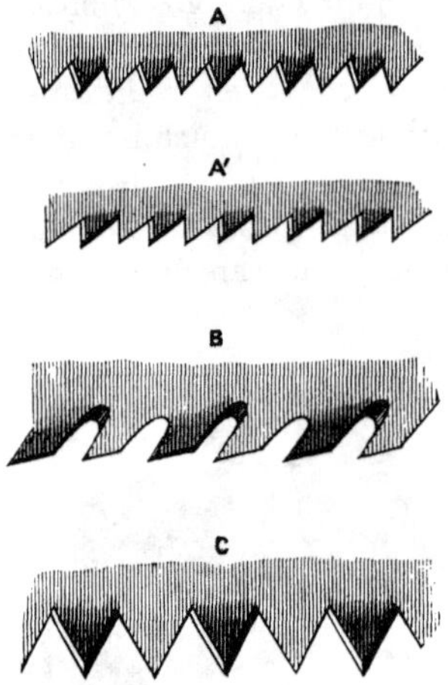

A

A′

B

C

FIG. 30.—Forms of Saw-Teeth. A Hand-saw teeth. A′ Ditto. B Gullet teeth. C Cross-cut teeth.

Each individual tooth may be not inaptly compared to a very narrow chisel, which being passed with a light scraping action many times in succession over the material produces a groove by the removal of successive portions of it. Or we may even go a step farther, and compare each tooth to the cutter of a plane, the duty of the stock being performed for it by all the adjacent teeth. For, the points of the teeth being (generally) in one straight line, they are compelled—like the plane-iron—to act first upon the projecting portions of the material, the undue penetration of each tooth being to a considerable extent prevented by the fact that its depth cannot exceed that of its neighbours. But this tendency is not entirely removed, as it is in the case of the plane; and it is therefore necessary—in making, for instance, a horizontal cut—that the vertical pressure (or what may be called the *feeding power*) shall be properly proportioned to the horizontal force. This proportion is determined by the weight of the saw-blade, and the position of the handle with respect to the line of the teeth, though it can of course be increased or diminished by the action of the wrist of the workman. Its amount must vary with each variation in the front and back angles of the teeth, the number of them in action at one time, the nature and direction of the fibres of the wood operated upon, &c., &c.; so that, considering how much difficulty its investigation would present, and how little profit, we shall do better, in this, as in many practical questions of its kind, to seek the guidance of experience, rather than to attempt to assume the leadership.

Much may also be learnt from the shapes and thicknesses of the various saw-blades, which will be found to be well

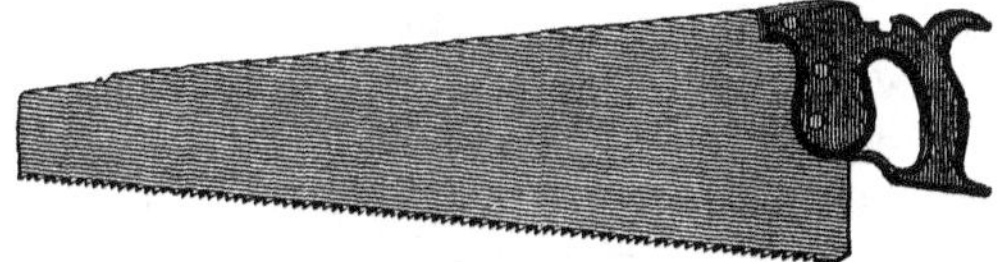

FIG. 31.—Hand-Saw (1/8).

suited to their respective requirements, though they have been arrived at without any aid from theorists. Thus the *hand-saw* (Fig. 31), which is intended for making straight cuts only, has considerable width given to its blade; but inasmuch as it must be able to bear sufficient thrust to make its cut without having an undue tendency to bend at any one point, it is made 'taper,' and not parallel throughout its length. For in the latter case the blade would be weakest near the handle. The thickness is also to some extent regulated by the same consideration; for, provided that the strength be sufficient, the thinner a saw-blade is made, the less material does it waste in making a cut, and the smaller is the amount of power which must be expended in the operation. Thus *compass-saws* (Fig. 32), and others which have narrow blades, so as to enable them to saw round curves of moderate radius, have much greater thickness given to them than is necessary or desirable for hand-saws,

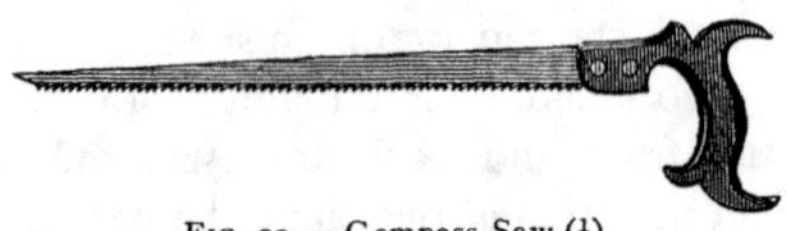

FIG. 32.— Compass-Saw (⅛).

so as to compensate for the diminished width, though this thickness is considerably reduced by grinding the back of the blade much more than the front in the process of manufacture, which is to some extent the practice with all taper saw-blades. From the position of the handle (Fig. 32), it is evident that it gives no feeding power whatever; that, in fact, pressure applied at the centre of the handle would force the teeth away from the work, instead of driving them into it. To compensate for this, the teeth of these saws are made of the form A′ (Fig. 30), for which less feeding force is required, the remaining deficiency being supplied by the wrist, assisted generally by the other hand of the operator.

Back-saws (Fig. 33), which are strengthened by a piece of iron or brass folded over the back of the blade, are able to

FIG. 33.—Back Saw (⅛).

have the thickness very much reduced. This, together with their great feeding power, in consequence of the handle being so much above the line of the teeth, causes them to work with great ease, their tendency being to cut with only too great avidity, which the wrist is often called upon to resist. The depth of the cut being limited to the width of the blade, it is necessary to make saws of this kind parallel instead of tapered.

By reversing the teeth of a saw, so as to make it cut

during the backward instead of the forward stroke, the thickness and width for enabling it to resist flexure is rendered unnecessary; for, whereas a thrust has a tendency to bend the blade, a pull only tends to straighten it. But with a few exceptions—such, for instance, as the saws with thin curved blades, which have now superseded the cumbrous double-toothed pruning-saws of former days—this principle is only applied in this country when the blade is stretched in some kind of frame. In all these *bow-saws* (such as Fig.

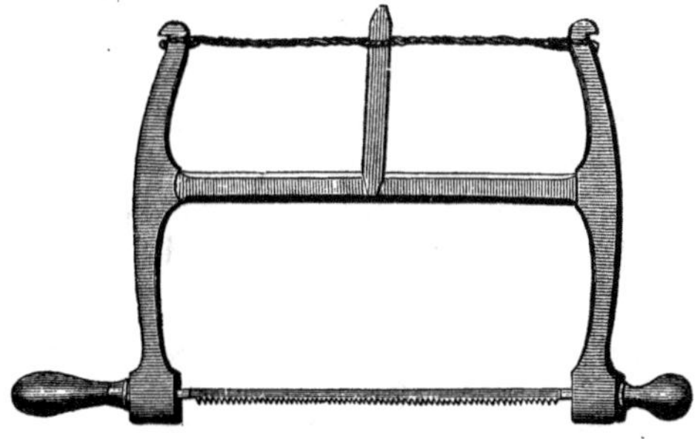

FIG. 34.—Bow-Saw (1/8).

34), the thrust is taken by the frame, and the pull or tension only by the blade, which can therefore be reduced, both in width and thickness, to any required extent. Saws of this kind are therefore used for sawing out all curves of small radius, for cutting metal, ivory, and other substances of which it is an object to remove as little as possible, and also in the endless-band sawing machines. Pit-saws and the smaller sizes of cross-cutting saws also frequently have frames, and on the continent of Europe and elsewhere frame saws are advantageously used for many purposes to which they are not applied in England.

Although the action of a saw is not to so great an extent dependent upon the keenness of its teeth, as is a plane upon the acuteness of its iron, occasional sharpening is of course necessary, and we shall therefore, as we have already done in the case of the plane, give a brief sketch of the operation, though it is doubtless a familiar one to many of our readers. For this purpose an ordinary grindstone clearly cannot be used, and although a saw-sharpening machine has of late years been introduced, which enables the gullet-teeth of circular and other large saws to be ground on an artificial emery wheel, it has not at present done much to supersede the saw-file in their case, whilst for angular teeth it is not in the least adapted.

For these, therefore—to which we shall confine our attention—triangular, or *three-square files*, are invariably employed, on which account the angle between the face of one tooth and the back of the next is in almost all cases the same as those of an equilateral triangle, or 60°. Their forms, and also those used for sharpening saws with gullet teeth, are given in Fig. 55. In order to apply them conveniently, the saw—having been previously removed from its frame, if it be a bow-saw—is fixed with its teeth upwards, in a vice, or a sawing-horse, the latter, however, being seldom or never used, except for pit-saws. But whatever the form of holder, the object to be attained is to support the saw blade between two jaws of wood or other soft material, at as small a distance from its teeth as can be done without interfering with the free use of the file. This, before being applied to the individual teeth, should be removed from its handle, and passed lengthwise a few times along the points of the whole of them—those at the ends, which come in for a smaller amount of wear, frequently requiring to be lowered or 'topped' more than the others, in order to bring them to the same level. By this treatment, a small horizontal facet will be left on the point of each tooth. In filing each of them separately—which is the next step—its newly-formed point must be brought exactly to the centre of this facet, which can be readily done by removing a larger proportion from the face or from the back of the tooth, as may be required. The file, however, should not be inserted into each space consecutively. When filing fine teeth it is advisable, and when filing coarse ones it is necessary, to do them alternately from opposite sides, and this is most easily effected by omitting every alternate space in the first instance, then releasing the saw from the vice, turning it end for end, and filing those previously omitted. The file is seldom held at right angles to the saw blade, but is inclined to it both horizontally and vertically to a greater or less degree, according to the softness or hardness of the timber upon which it is to be used, the vertical inclination being about one half of the horizontal, as the diagram (Fig. 35) shows. Bearing in mind that (in filing angular teeth) the handle of the file should never be raised above its point —in other words, that the stroke should be upwards instead of downwards, and that the greater the noise the less is the effect produced upon the saw — the reader will have no difficulty in determining for himself which are the spaces which should belong to the first series and which to the second.

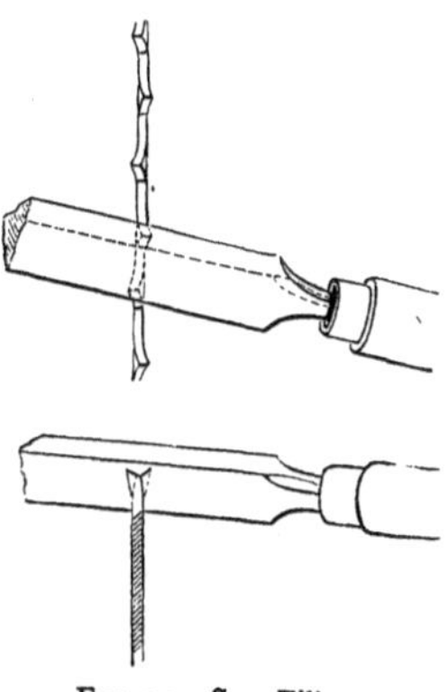

FIG. 35.—Saw-Filing.

But after the filing has been completed, the performance of a soft-wood saw will not be found to be satisfactory if attention be not also paid to *setting* it. This consists in slightly bending the teeth alternately to the right hand and to the left, to an extent which varies with the nature of the substance to be cut. The effect of it is to make the width of the cut, or *kerf*, rather greater than the thickness of the saw blade, which thus encounters little or none of the resistance from 'binding,' which would result if both the blade and the cut were of exactly the same width. Fig. 36 shows the back view of a saw, by which a workman examines the amount and the evenness of its set; and also a portion of he same in section.

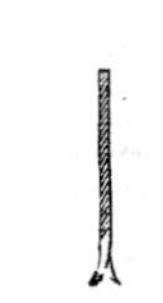

FIG. 36.
Set of Saw-Teeth.

There are various ways in which the set can be given to the teeth, of which that almost invariably employed by the saw-maker is undoubtedly the best in skilful hands. This consists in laying the toothed edge of the saw-blade upon a small stake, or anvil, with a long and sharply-curved surface, which when in use is fixed in an ordinary vice. One or two blows from a light, narrow-faced hammer then suffices to bend each alternate tooth downwards, to an extent depending upon the curvature of the anvil, and the position of the tooth upon it. When half the teeth have been done in this way, the saw is turned end for end, and the other half is similarly treated.

Joiners and others who are users, but not makers, of saws, generally employ some kind of *saw-set* in preference to the saw-setting hammer. In its most simple form a saw-set may be merely a piece of steel plate of a thickness not exceeding the width of the saw-tooth, having in its edge a parallel-sided notch, equal in width to the thickness of the blade. For convenience, however, a saw-set is usually provided with a number of notches of varying widths, so as to enable it to be used for saw-blades of different thicknesses. The saw to be operated upon is fixed with its teeth upwards, in the same manner as for filing, and the teeth, separately inserted a short distance into a notch of suitable width, are alternately bent in opposite directions by raising or lowering the handle of the saw-set. To do this equally to all the teeth without any kind of stop or guide is by no means an easy task, and since uniformity of set is of great importance, some less simple form of saw-set is generally required to produce satisfactory results in inexperienced hands. The addition of an adjustable stop to the notched form of the instrument just described, is of some assistance, as it prevents the teeth from being bent to different angles; but those known as *patent saw-sets*, by which the teeth are clipped between the jaws of a pair of pliers of peculiar form, are provided with adjustments which leave nothing to the discretion or the indiscretion of the operator. In others, again, a kind of miniature swage is held up by a spring over a small stake or anvil, upon which the saw-block rests. A single blow from a hammer at the top of the swage completes the set of each tooth, perfect uniformity being ensured by adjustments which regulate the inclination of the blade and the distance to which it can extend on to the anvil. These saw-sets, if made more generally applicable to saws of various shapes and sizes than they are in their present form, would probably be found to be more speedy and effectual than any others.

As a test of the uniformity with which the operations of setting and sharpening a saw has been conducted, it has been suggested to place a needle in the angular groove formed by the tops of the teeth (which appears in the sections, both in Fig. 35 and Fig. 36). On gently raising one end of the saw, the needle should run from end to end without leaving the groove.

Although our mention of the setting of saws has been deferred till after the subject of filing them had been dismissed, this order is not that which should be followed in the process of sharpening. If indeed so large an amount of topping be required as almost to amount to re-forming the teeth, the course adopted in the original manufacture—in which they are filed to their true form, set, and then again slightly filed to sharpen them—may of course be advantageously employed; but in ordinary cases, setting should precede filing. The 'burr' thrown up by the file, which has much to do with the keenness of a newly-sharpened saw, is then retained uninjured. In fret-cutting and some other kinds of saws this burr constitutes their only set; but as the prevalent practice is to file such teeth from one side only, instead of alternately from opposite sides, these saws have the tendency to 'draw,' instead of making a straight cut, which is always the result of giving more set to one side than to the other.

In concluding our notice of this invaluable instrument, we subjoin a few particulars of the saws which are most likely to come under the notice of the reader, extracted partly from the second volume of Mr. Holtzapffel's work, already referred to. (See table on next page.)

Of the wood-tools generally used for *boring*, some of which next demand our attention, the *brad-awl* (Fig. 37)

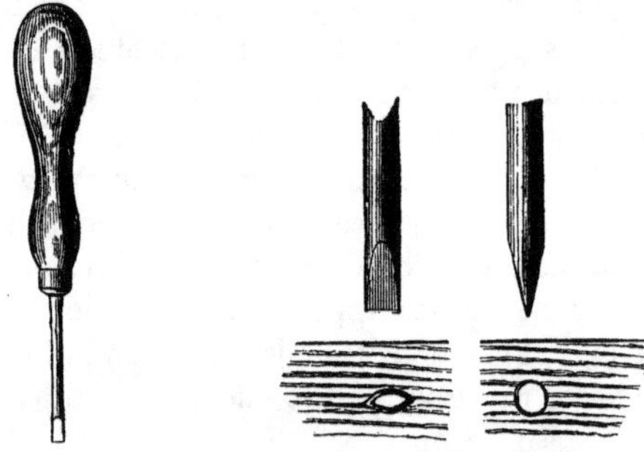

FIG. 37.—Brad-awl (¼). FIG. 38.—Action of Brad-awl.

seems to claim precedence; not, indeed, from the perfection of its performance, but because, being very simple in its form and speedy in its action, it enjoys a sort of monopoly wherever small holes have to be made in soft wood. When properly placed, with its edge across, and not in the direction of the fibres, so that it may cut them instead of merely forcing them asunder, a sharp brad-awl—especially if driven with a hammer or mallet instead of being

TAPER SAWS.

	Form of tooth[1]	Length of blade	Width Max.	Width Min.	No. of teeth per inch	Thickness: By Birmingham Wire Gauge	Thickness: By Whitworth's Standard Gauge[3]
With a handle at each end		ft.	ins.	ins.			
Cross-cut saw[2] . . .	C	4—10	6 to 12	3 7	8 in 5 ins.	12 to 15	110—70
Pit-saw	B	4—10	9 to 12	3½ 5	4 in 3 ,,	12 to 16	110—65
Pit frame-saw	B	4—10	7 to 11	3 4½	8 in 5 ,,	15 to 18	70—50
With a handle at one end.		in.					
Rip-saw . . . . . Half rip-saw . . .	A or A′	28—30	8 to 9	3 4	3— 3½	18 or 19	50—40
Hand-saw	A or A′	10—26	5 to 7½	2½ 3	5— 6	18 or 19	50—40
Panel-saw . . . . Fine panel-saw . .	A or A′	10—26	4 to 7½	2½ 2	7—12	19 or 20	40—36
Table-saw	A or A′	18—26	1¾ to 2¼	1 1½	7— 8	16 to 19	65—40
Compass-saw	A′	10—20	1 to 1½	½ ¾	7— 9	18 or 19	50—40
Keyhole-saw	A′	6—12			9—10	19 or 20	40—36
PARALLEL SAWS.							
With backs.		in.					
Tenon-saw	A or A′	12—16	3 —4		10	21 to 23	32—24
Dovetail-saw	A or A′	6—10	1½—2		14—18	24 or 25	22—20
With frames.							
Woodcutter's or billet saw	A or C	24—36	2 —3½		3— 4	19 to 22	40—28
Turning, or sweep-saw .	A′	6—22	$\frac{1}{10}$— ⅝		10—20	19 to 24	40—22
Smith's frame-saw . .	A′	6—18	¼— ⅞		10—14	20 to 24	36—22
Fret-saw	A′	3—14	$\frac{1}{25}$—$\frac{1}{60}$		30—90	$\frac{1}{50}$ to $\frac{1}{200}$ in.	20— 5

worked by hand—will produce in deal and other loose-fibred woods a fairly clean and circular hole, without much

[1] See Fig. 30. The forms there given are general examples only, and not accurate representations of either the sizes or the angles of the teeth.

[2] Fish-bellied, having its greatest width at the centre of the blade. Teeth somewhat resembling B, and various others are also used.

[3] Expressing the thickness in thousandths of an inch. See p. 24.

danger of splitting, as Fig. 38 will serve to explain. For boring hard woods, which require a portion of their substance to be removed, brad-awls are not adapted, as they have no means of effecting this. Recourse is then generally had to the *plain gimlet*, or to the *twisted gimlet* (Fig. 39), although in this latter the strength is insufficient for boring woods of which the hardness is excessive. The size of gimlets like those shown in the figures does not in general exceed $\frac{3}{8}$ of an inch; others, however, resembling the screw-auger (Fig. 41) are as much as $\frac{3}{4}$ of an inch. The pointed screw by which they are made self-feeding, has considerable tendency to split any wood that is given to splitting, and the insufficient size of the groove to contain the whole of the mate-

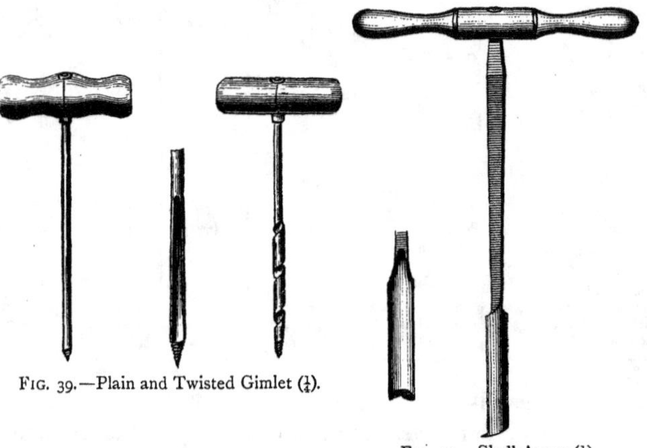

FIG. 39.—Plain and Twisted Gimlet ($\frac{1}{4}$).

FIG. 40.—Shell Auger ($\frac{1}{8}$).

rial displaced causes a gimlet soon to become choked. This necessitates its frequent withdrawal when boring a deep hole, whereby its naturally slow action is rendered slower. At best, indeed, this tool is far from perfect, the expenditure of both time and power being out of all proportion to the effects produced. But it is perhaps to these imperfections that we owe the introduction of pointed wood screws, which are a very great improvement on the old-fashioned ones with blunt ends.

The principal tool for boring holes, of which the diameter or the depth is too great for a gimlet, is the *auger*. The *shell-auger* and also the *screw-auger*—which has to a great extent superseded it for boring soft wood—are shown in Figs. 40 and 41. But for hard woods their great strength is likely to enable shell-augers to hold their own, in spite of the advantages which their rivals possess in being self-feeding, capable of being worked with much less power, and not so liable to become choked. The cutting ends of a shell-auger and two kinds of screw-auger are shown enlarged in these figures. For fixing the long cross handle, by which the requisite leverage for working an auger is obtained, the upper end of the stem is either drawn out to a tang of great width but slight thickness, which can be driven through the centre of the handle and clenched; or an eye is formed, through which it can be passed. In the engravings the shell-auger (Fig. 40) is 'tanged,' and the screw-auger (Fig. 41) is 'eyed;' the tang being always placed with its greatest width across the grain of the wood, so as to reduce the risk of splitting it. Occasionally the tang is made square, like that of a brace-bitt, which enables one handle to serve for several augers. The diameter of the cutting part of an auger

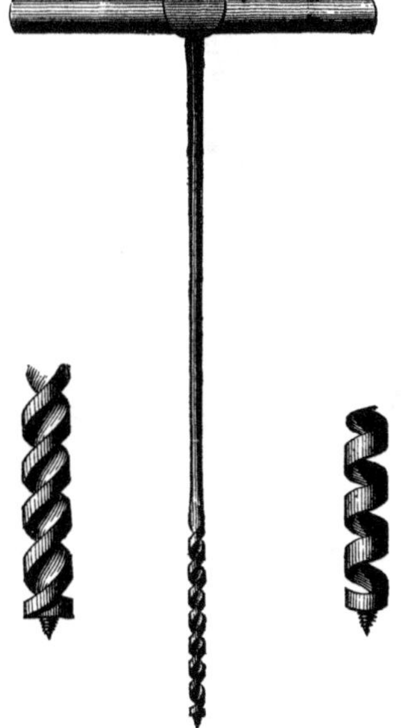

FIG. 41.—Screw-Augers (⅛).

generally ranges from $\frac{3}{8}$ of an inch to 2 or $2\frac{1}{2}$ inches. Shell-augers of 3 inches diameter and upwards were formerly used for boring wooden pumps and pipes, but with them they have now ceased to be manufactured. Shell-augers when once started—an operation in which they require the assistance of a gouge or other tool—are kept straight in their course by the guiding power of the hollow parallel 'shell,' which extends for some inches above the cutter. But the friction thus produced consumes so much power, especially in the larger sizes, that it is more advantageous to bore a hole of large diameter in two operations; first making a small hole, truly concentric with that required, and afterwards enlarging it by means of a tool on the same principle as the plug-bitt shown in Fig. 45. This, however, does not apply to those screw-augers which have two cutting edges, the resistance encountered by one edge being equal to that met with by the other, so that the whole tool is in a state of equilibrium, and little or no guiding is required. We shall revert to this subject in connection with the drill (see Fig. 60).

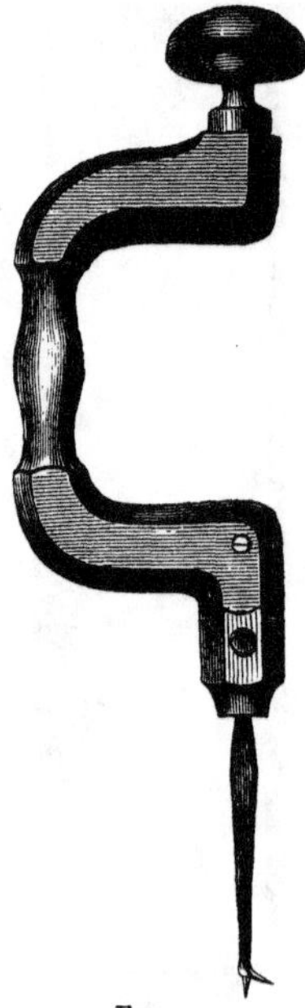
Fig. 42.
Brace and Bitt ($\frac{1}{8}$).

None of the preceding tools are well adapted for fine work, in which it is essential that holes should be perfectly smooth and accurate. For these purposes a *brace* and set of *bitts* offer great advantages, not the least of which is the number and variety of borers which can thus be carried in a small space. The English pattern of *brace* (or *stock*) is shown in Fig. 42. It is usually accompanied by a set of from one to three dozen bitts, of which the following are the principal ones :—

Quill-bitts (Fig. 43), for boring across the grain of the wood only, which they do both well and quickly, cutting the fibres instead of tearing them apart. These, and almost all other brace-bitts, are fed by the pressure of the chest of the workman against the top of the brace.

Nose-bitts (Fig. 44) differ only from quill-bitts in having at their extremity a transverse cutter, which enables them to be used for boring in the direction of the grain. They much resemble shell-augers.

Centre-bitts (Fig. 45), for boring holes of small depth (up to about two inches in diameter) at a single operation, or rather by two distinct operations which are carried on simultaneously, consist of a triangular pin or 'centre'

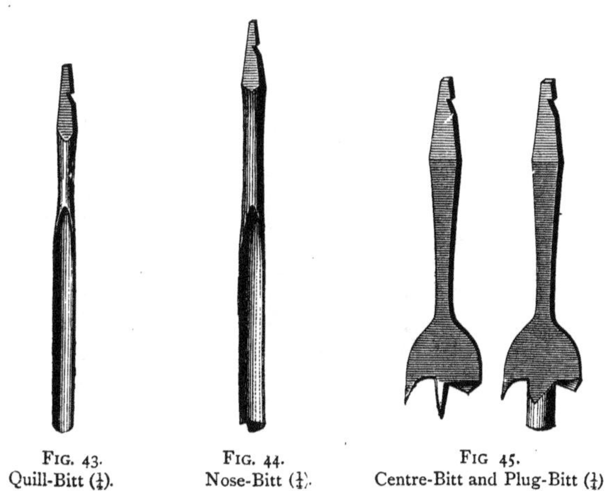

FIG. 43. Quill-Bitt (¼). FIG. 44. Nose-Bitt (¼). FIG 45. Centre-Bitt and Plug-Bitt (¼)

carrying on one side a knife-edge or 'nicker,' and on the opposite side a 'cutter.' The distance between the centre and the outer face of the nicker is exactly equal to the radius of the hole produced, the width of the cutter being rather less than this. At each revolution the nicker makes a clean circular incision, and from the space enclosed by it the cutter then removes a helical shaving, whose thickness depends upon the pressure applied to it. The efficiency of this tool is much greater when it is used across the grain of the wood than in the direction of it. For boring a large hole concentric with an existing small one a centre-bitt is not adapted, unless the central pin be enlarged so as completely to fill the hole which is to be followed. It is then called a *plug-bitt*, but it is a tool of more special than general application.

Besides those above mentioned, a set of bitts generally contains several *taper-bitts* or *rimers* (of which Fig. 46 shows that used for wood), *countersinks*, for enlarging the entrance of holes to admit the heads of screws, of which Fig. 47 shows two kinds, and one or more *screwdrivers* for use with

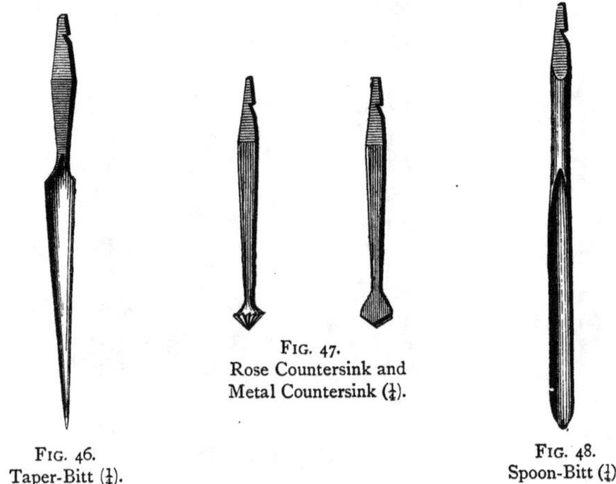

FIG. 47.
Rose Countersink and Metal Countersink (¼).

FIG. 46.
Taper-Bitt (¼).

FIG. 48.
Spoon-Bitt (¼).

the brace. Many other bitts are also much used, though not included in ordinary sets. Such are *spiral-bitts* (resembling Fig. 41); *spoon-bitts* (Fig. 48), and various kinds of *expanding centre-bitts*; but to describe, or even to enumerate the boring and other cutting tools which are employed in the various special wood-working trades in this country, would carry us very far beyond our limits. The small number of which we have been able to give figures or descriptions have been selected as being good examples of the various types, which—with but one or two exceptions—are of common occurrence.

But we are unwilling to take our leave of wood-tools without reverting to the subjects of grinding and setting their cutting edges, necessary operations which should be among the very first lessons of the beginner, and which cannot be neglected with impunity by the oldest and most skilful hand.

On attempting, for instance, for the first time to sharpen his plane-iron in the manner already described, the reader will almost certainly find himself beset with difficulties. Assuming that his grindstone runs perfectly true and that its face is neither 'hollow' nor 'round,' the selection and maintenance of the proper angle at which to hold the tool will be found to be far from an easy matter, any variation either in its height or in the angle at which it is applied to the stone causing a fresh facet to be formed upon it. The tendency is thus to grind the bevel convex, and this must be carefully resisted by obliging it constantly to keep the same position. When this has been successfully accomplished, the facet will be slightly and uniformly concave to an extent which will vary with the diameter of the grindstone. Attention must also be paid to forming the edge of the tool square to its sides, by keeping every part of it evenly pressed against the stone; which is most easily done by placing the fingers of the left hand on its upper surface, as nearly as possible over the centre of the bevel which is being ground upon its lower one. The amount of pressure which must be applied will of course vary much with the width of the tool, and it will also be influenced by the direction in which the grindstone is revolving, more being required when its upper portion moves away from the operator, and less when it runs towards him. In the latter case, however, the edge of the tool is more liable to catch, and thereby to damage both itself and the face of the stone; whilst in the former a wire edge somewhat difficult of removal is thrown up directly the grinding has been continued sufficiently long to obliterate the small facet previously left by the oilstone. An obvious remedy for this is to discontinue the operation before the grindstone quite reaches the cutting edge, whenever the removal of notches or other inequalities does not necessitate its continuance. In the case, therefore, of such tools as after being ground are to be finished upon an oilstone, the direction in which the stone revolves may be said to be immaterial; for others—such as turning tools for metal—of which the cutting edges are wholly produced upon the grindstone, it is preferable that it should be driven towards the operator.

Since the difficulty of grinding edge-tools properly is enormously increased if the grindstone be allowed to lose its true cylindrical form, or if its face be worn unequally, care should always be taken to guard against this by giving a slow lateral motion to a tool whilst it is being ground. Neglect of this in the case of those which are flat, but of small width—such as chisels, &c.—soon results in the formation of a hollow at the part of the stone at which the work is chiefly performed, which causes it to give a curved instead of a straight edge to those—such as plane-irons—of which the width is greater. These indeed cannot possibly be ground correctly, if the centre of the grindstone face be worn hollow to any considerable extent.

In grinding carpenters' gouges and other tools of which the bevelled edge is curved, this traversing motion is even more necessary, for if held constantly in the same position they very rapidly 'score' the face of the stone and make the commencement of a groove, which is not easily got rid of. Where much use exists for gouges or any kind of tool of which the convex surface requires to be frequently ground, it is preferable to devote a separate stone to them and to allow its face to become grooved. When their edges, although curved in one direction, are straight and at right angles to their length, the required curvature can be easily given by constantly turning the wrist of the right hand, in which the tool is grasped, backwards and forwards through a larger or smaller arc of a circle according to the amount of the curvature. Gouges, however, are ground occasionally on their concave instead of on their convex sides, and for these as well as for various moulding and other tools, grindstones, if used at all, must have corresponding projections turned upon their faces. But in such cases—as also in sharpening the generality of boring tools, &c.—the grindstone is frequently abandoned in favour of the file ; or recourse is had either to an iron or copper *lap* charged with emery powder, or to the same material in the consolidated form of an *emery wheel.*

In order to prevent the temper of edge-tools from being injured by the heat generated in grinding them, and also to clear the grain of a grindstone from the detached particles of sand and steel, a constant supply of water is required. Small stones, of which the rapid revolution causes any excess to be thrown off by centrifugal force, are most conveniently supplied by allowing a small quantity to drip upon their faces, whilst those of larger diameter are generally allowed to dip into some water contained in a trough placed below them. The chief disadvantage of this arrangement is the temptation which exists to leave the stone standing in the water whilst at rest, of which the result is to soften it and to cause it to wear unequally. If at any time from this or from other causes a grindstone betrays any eccentricity, the process of turning up, though not perhaps an agreeable one, should at once be performed. If deferred, the evil will always be found to increase, until at last it becomes intolerable, and 'hacking' (*i.e.* dressing off the more prominent parts with a hack hammer) must be resorted to. It is indeed the best method of correcting large grindstones, whether much or little has to be removed from their face; but for small ones it is only necessary in extreme cases, in which it is preliminary to, but not a substitute for turning. For this latter purpose the stone should be dry and should be driven slowly, the tool, which is generally a worn-out three-square file, being held with its point slightly inclined downwards upon a rest fixed as closely as possible to the face of the grindstone. The triangular section of the tool enables a fresh edge to be brought into play by merely turning it over as soon as its upper edge ceases to cut.

Of the various kinds of grit-stone which are used for grinding, that known as *Bilston* (from the place of that name in Staffordshire, where it is extensively quarried) is by far the best suited for sharpening the generality of edge-tools, having a fine and quick-cutting grain without too great hardness. Others obtained from Northumberland, Yorkshire, Derbyshire, &c., are much used in the manufacture of cutlery and hardware, but for the most part (although they vary considerably,) they are of coarser grain, and consequently require to be driven at a much higher speed. For grinding the concave facets of moulding tools, and for some special purposes for which great smoothness of grain is required, thin discs of a good but costly kind of *Bohemian* stone are employed. Recently Mr. Frederick Ransome has turned his attention to the manufacture of grindstones by the artificial process with which his name is connected. These are reported as having certain advantages over the natural stone, and amongst others that of a reduction in the first cost of the grindstone.

The hones and oilstones available for setting the edges of cutting-tools form a much larger class than the stones used for grinding them. But although these are obtained from many parts of the globe they for the most part occur either within very narrow limits or in comparatively small pieces—

this indeed being more especially true of some of the best varieties. On this account they are rarely mounted and used in the same manner as grindstones, although if they were obtainable in equal abundance this would probably be the general practice. As it is, the workman is obliged to content himself with a rectangular piece from five to nine inches long, upon the surface of which he produces a small facet at the extreme edge of his plane-iron or other tool, by rubbing it briskly backwards and forwards as nearly as possible along its entire length. To prevent an oilstone from getting broken—an accident to which some kinds are very liable—and to protect its oily surface from dust, it should always be mounted in a piece of hard wood and be provided with a cover.

In theory the operation of setting a cutting-tool—like that of grinding it—is simplicity itself; being merely the formation of two flat facets inclined to one another at a certain angle, their intersection forming the cutting edge; and indeed in tools which are ground with a single bevel it is only necessary to produce one of these facets, the flat unground side taking the place of the second. But in practice there is great difficulty in continuing the operation just so long as to obtain the complete intersection of the facets at all parts of the edge, without anywhere throwing up the wiry film already mentioned, which is formed upon the oilstone as well as upon the grindstone. Its presence, however, is quite incompatible with the possession of a keen and durable edge, and its removal must therefore be effected whenever it occurs; which may be done, though not very readily, by drawing the edge across a piece of soft wood, or over the thumb nail. In any tool ground with a single bevel, slight treatment of the unground side also upon the oilstone will be found to assist in removing the wire edge, but in doing this care must be taken to keep it flat on the stone and in no case to form a facet on its edge, which, as already pointed out, in almost all cases is a certain method of destroying its efficiency. Chisels and other straight-edged tools, when being set, are best held in such a manner that the direction of the motion of the hands is nearly but not quite at right angles to the line of the cutting-edges, as may be seen in the operation of setting a plane-iron, which is represented in the engraving, Fig. 49. But in the case of carpenters' gouges—which we have before taken as the type of edge-tools with this kind of curvature—it is preferable to set the edge by moving it in the direction of its length. For

FIG. 49.—Setting a Plane-iron.

this purpose the right hand, in which the tool is grasped, is held considerably to one side of the stone, every portion of the edge being then brought into contact with its surface at each forward or backward stroke by means of a similar wrist motion to that given in grinding it. Concave facets, which of course cannot be produced upon a flat stone, are formed by means of *oil slips*, which are merely thin pieces of oilstone of which the edges are rounded with the required curvature.

The chief points which require to be attended to in connection with an oilstone are: 1st, the maintenance of its flatness, which it always loses with use, owing to greater wear taking place at the centre than at the ends. It must, therefore, from time to time be surfaced by grinding it down on the side of a grindstone, or with fine sand and water upon a flat stone or metal surface. Secondly, its cutting power must be preserved; the coagulation of the oil upon its surface frequently either wholly destroying or greatly interfering with the abrading action of the particles of silica, to which both grit-stones and oilstones owe their property of acting upon hardened steel. For this the best remedy is to rub the face occasionally with a lump of moistened pumice-stone. In every well-regulated workshop, both grindstone and oilstone should be constantly kept in a state of efficiency, and be ready for use at any moment.

The following are the principal kinds of hone and oilstone from which a workman is likely to be able to make his selection—the order in which they are placed being approximately that of their abrading power; those at the top of the list being the "fastest cutting," a quality which is generally accompanied by a want of fineness in the edge produced.

1. Washita Oilstone.—A very compact white sandstone, of rather recent introduction, almost resembling Carrara marble in appearance. Although it does not greatly differ in price from Turkey stone, its much greater uniformity and slightly more rapid cutting property cause it to be in more favour with carpenters and others, with whom coarseness of edge is not an objection.

2. Turkey Oilstone.—When of good quality no better substance can be employed for setting tools for which great fineness of edge is not required, since it cuts the hardest

steel with avidity even when but little pressure is applied. At the same time it is of a close grain, and is not easily scratched. Unfortunately it is very variable in quality as it is also in colour; the latter, which is called 'white,' 'grey,' or 'black,' being generally a veined mixture of different shades of bluish and brownish greys. Its cost is about three times that of the stone next mentioned.

3. Charnley Forest Stone.—Found near Mount Sorrel, in Leicestershire. This is the best of our native oilstones, and has long been a favourite with carpenters and others, that from the Whittle Hill Quarries, which is of a grey colour, dappled or streaked with red, being considered to be the best. Till lately this has hardly been obtainable—the only representative of Charnley Forest stone being a rather inferior one with a decided green tint. Both of them, however, give a very fairly fine edge, but do not cut quite so rapidly or with as slight pressure as Turkey.

4. Canada Oilstone.—A very fine porous sandstone of a greyish white colour, which has been recently introduced. Being much less compact than any of the preceding stones, it is much more rapidly worn away. Its first cost is, however, rather less than that of Charnley Forest.

5. Grecian Hone.—Under this name a slaty stone is imported, which is of a greenish colour, and although said to be superior to Welsh oilstone, does not greatly differ from it in appearance.

6. Welsh Oilstone.—A hardish stone of a green colour and slaty texture, inferior to the Charnley Forest for joiners' use. In price it is about the same, as also is the Grecian hone, No. 5.

7. Arkansas Oilstone.—Cuts slowly, but is very superior to all those above mentioned for giving a fine edge to surgical instruments, &c. Although extremely costly—its price being about four times that of Turkey stone—it is extensively used for such purposes. In colour it resembles

Washita oilstone, but it is of very much finer grain and wears away very slowly.

8. German Hone.—Thin slabs of a very soft yellow stone, cemented upon a rather harder but similar material of a slate-blue colour, are imported and sold under this name. The extreme softness of the former renders it almost useless for such edge-tools as we have been considering, although it is well adapted for setting razors, to which it imparts an edge of great smoothness and delicacy.

Besides the above, some other kinds of stone, which require to be used with water instead of with oil, may be employed for sharpening edge-tools. Such is the Water of Ayr Stone, which requires to be kept constantly moist to prevent it from becoming hard. Its chief use, however, is for smoothing marble, copper, &c.

The oil applied to an oilstone should have the same quality of resisting the action of the air as that used for the lubrication of machinery; sperm, neat's-foot, and olive oil being well adapted for it.

Before closing this chapter we should mention one other wood-tool which differs from all the others both in the form of its edge and in the treatment by which it is produced. This is the *wood scraper*, which consists, as appears in Fig. 50, of a small rectangular steel plate, clipped upon one of its longer

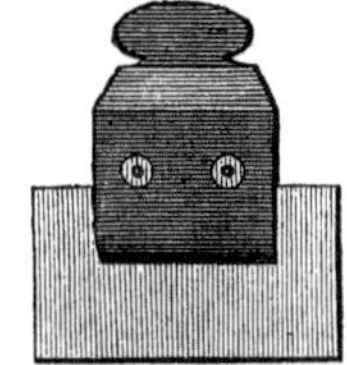

FIG. 50.—Wood Scraper (⅙).

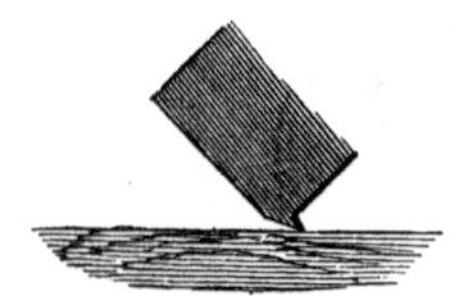

FIG. 51.

sides by a handle of peculiar form, which is easily removable. Its edge, a section of which is shown much enlarged in Fig. 51, is first brought exactly to a right angle by being set upon an oilstone (it then resembles the upper angle in the diagram); after which a slight and uniform 'burr' is thrown up by passing along it the rounded back of a gouge, the stem of a large brad-awl, or any similar piece of hardened steel.

A glance at the figure will explain the action of this burr, which, in fact, forms an actual and somewhat delicate cutting edge, which is very effective in giving a smooth surface to hard and cross-grained wood. The natural fracture of a piece of window glass, which was formerly used for scraping, and in which, of course, no burr can exist, is much less satisfactory for this purpose, even in cases in which the impossibility of thus obtaining a really straight edge may not be an objection.

But when ground, sifted, and glued upon paper or cloth, glass is a most valuable assistant to the worker in wood. With *glass-paper* or *glass-cloth* the final smoothing of the more finished kinds of wood-work is almost always performed, the method of applying it to flat surfaces generally being to fold it over a piece of cork with a flat rectangular face; and similar means applicable to curved surfaces will readily suggest themselves. To the turner and the polisher glass-paper is almost indispensable—the following being the sizes generally used in London:—No. 00 or 'Flour' (which is the finest); No. 0; No. 1; No. 1½; Fine 2 (or F 2); Middle 2 (or M 2); Strong 2 (or S 2); No. 2½; and No. 3 (which is the coarsest).

CHAPTER III.

ON HAND-TOOLS USED FOR METAL.

HAND-TOOLS such as those of which we have been considering the forms and the modes of action, although they may be employed with great effect in the treatment of timber and other materials which do not greatly exceed it in hardness, would be powerless to perform the more severe labour of cutting and shaping cast or wrought iron or steel—operations which form by far the larger portion of the practice in engineering workshops. In all the more arduous of these, indeed, the machinist avails himself of the various machine-tools to which we have already alluded—many of which will be subsequently described in detail—but before their invention and introduction the comparatively small amount of working in the harder metals which was done at all, was performed with the aid of hand-tools only. These for the most part are still retained, though their duties lie within a smaller compass; we shall, therefore, devote the present chapter to noticing them in a slightly less cursory manner than that in which we have been compelled to dismiss the subject of wood-tools—their much smaller number rendering this possible.

One great distinguishing feature of all cutting-tools for iron and the harder metals is the much greater thickness which is given to their edges; their cutting angles being from 50° to as much as 90°; whereas those of wood-tools, as we have seen, rarely exceed the smaller of these angles, and are for the most part much below it. An example of this may be seen in the various forms of chipping chisel, of which the edges when intended to be used upon wrought iron are ground to an angle of 50°; and when cast iron has to be operated upon, to an angle of 60°; although thin edges, if they could be made to stand (which they cannot), would undoubtedly require a smaller amount of power to be expended in making a cut in both cases.

In working the harder metals by hand, the *chipping chisel*, driven by a succession of blows from a *hand hammer* (ordinary forms of both of which are shown in Fig. 52), constitutes a much more important tool, and one which is in much more frequent requisition than its representative among wood-tools; the hard and stubborn nature of these materials entirely forbidding the use of any cutting-tool corresponding to the axe, with which the first roughing-out of timber is effected; and although it might be thought that by careful forging and casting the desired forms could be produced with such accuracy that no after-treatment except filing should be necessary—and this is in great measure true of forgings—it is far from being the case with castings. And here it should be noted that throughout these pages, when forgings and castings are spoken of, they refer to works in wrought and cast *iron* only, except in the rare cases in which other metals are specified. Castings then—which are produced by pouring melted iron into sand moulds—always have their surfaces hardened by the chilling effect of the moulds, and in addition are more or less covered with obstinately adherent particles of sand. These causes combine to give them a hard 'skin,' against which the teeth of a file are almost powerless, so that every part of a casting which is to be filed up, requires, as a preliminary step, the removal of this skin, or at least of the sand which it con-

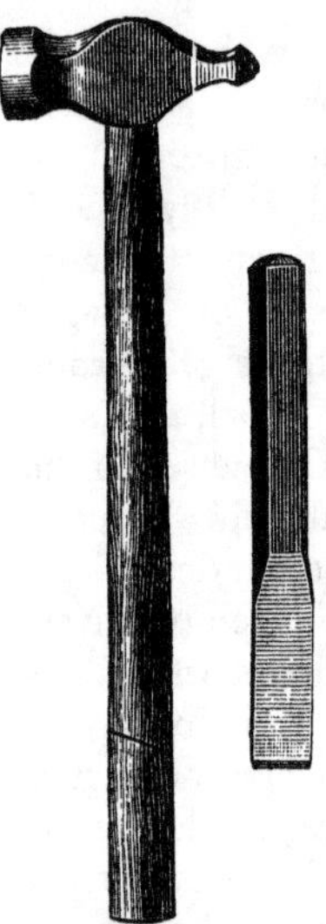

FIG. 52.—Hammer and Chipping Chisel (¼).

tains. This is generally done by cutting away the whole surface to the depth of about $\frac{1}{32}$ of an inch with a chipping chisel of about the size and width of edge represented above (*i.e.* from 6 to 8 inches long, and from $\frac{3}{4}$ to 1 inch wide); but in the treatment of large surfaces the operation is much facilitated by first making a series of parallel grooves with a *cross-cut chisel*, of which Fig. 53 is an example, though its edge may even be much narrower with advantage. Its width regulates that of the grooves, and their distance apart should not exceed the width of the chipping chisel, with which the intermediate portions of the material are to be removed. When small V-shaped grooves have to be cut with a chisel, its shape is modified accordingly; the single bevel with which it is ground presenting the form of a 'lozenge,' or 'diamond'; on this account it has received the name of a *Diamond-point.* The weight of the hand hammer ordinarily used with a chipping chisel varies from one to two pounds—its form also varying according to the fancy of the workman.

FIG. 53.—Cross-cut Chisel (¼).

The depth to which the edge of a chisel penetrates is regulated by the strength of the blow and the angle at which it is held. Since, in the choice and maintenance of the latter, the angles of the double-bevelled edge render little or no assistance—in other words, since it possesses so very small an amount of guiding power—it is not to be wondered at that proficiency in the art of using a chisel can only be acquired by practice. Among other points a beginner should remember :—First, that the edge of a chisel may be slightly rounded with advantage, so that its corners may not have a tendency to score the work or be easily broken off; secondly, that in operating upon any surface of which the width exceeds that of the chisel, the line of advance should for the same reason be kept constantly convex; and thirdly, that if he desires to avoid collisions between his hammer and his knuckles, he will do well to keep its face and the end of his chisel free from grease.

For large castings, or for portions of them from which a considerable quantity of metal has to be removed, *Flogging*

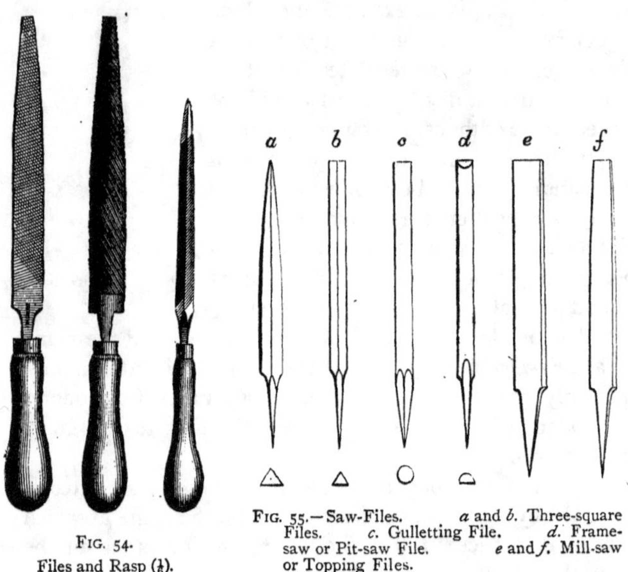

FIG. 54.
Files and Rasp ($\frac{1}{6}$).

FIG. 55.—Saw-Files. *a* and *b*. Three-square Files. *c*. Gulletting File. *d*. Frame-saw or Pit-saw File. *e* and *f*. Mill-saw or Topping Files.

Chisels, which are held in hazel rods or 'withes' by one man and struck with a sledge hammer by another, are substituted for the ordinary hand chipping chisels. But in many cases the skin of a casting is removed by grinding, or by 'pickling' it with dilute sulphuric acid, which dissolves some of the iron and thereby releases the particles of sand; either of which methods renders unnecessary the tedious and expensive process of chipping.

We next come to the consideration of *Files*, which in their almost endless variety constitute by far the most indispensable of all the cutting-tools which belong to the present chapter. In technical language, indeed, they are not comprehended in this term, any more than saws are included amongst the edge-tools used for wood; but we cannot see that any good purpose would be served by excluding either the one or the other from the positions which they now occupy in the present work. Although the use of files is by no means confined to the working of metals, some kinds—as we shall see—being but little adapted for the treatment of hard materials, yet this constitutes by far the larger proportion of their duty. On this account we have hitherto omitted all mention of them, preferring to group together the whole of those which can be noticed within the narrow limits at our disposal.

For the complete description of almost every individual file, five different properties must be stated :—First, its length ; secondly, its contour ; thirdly, the form of its cross section ; fourthly, the kind of 'cut' by which its teeth are formed ; and lastly, its degree of fineness. The examples represented in Fig. 54 will explain this ; the first of them being a 9-inch, taper, flat, bastard, double-cut file ; the second a 9-inch taper, half-round smooth rasp ; and the third a 6-inch, taper, three-square, second single-cut saw-file. With these characteristics we will deal seriatim.

The *length* of a file is measured in inches from the shoulder to the extremity, the tang or point for driving into its handle not being included.

The contour is described by the terms *parallel* (or *blunt*), and *taper* (or *pointed*), which explain themselves. But 'parallel' includes all files of which the width is tolerably uniform throughout, although the thickness may be considerably greater near the centre than at the ends. Indeed very few files have their section really equal throughout their length. Files are sometimes said to be more or less *bellied* according as they are much or only slightly tapered. In the latter case they may also be described as *blunt-pointed.* Examples of both taper and parallel files will be found in Fig. 55, which gives the outline of those generally used for sharpening saws.

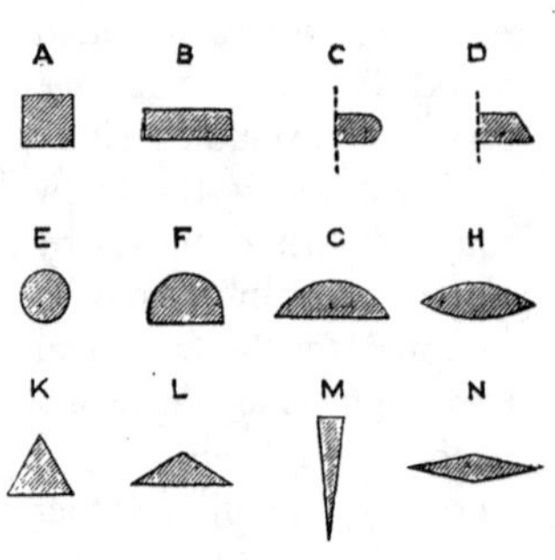

FIG. 56.—Sections of Files.

In Fig. 56 types of most of the *file sections* in common use are given. They are known by the following names :—

A. *Square files,* which are made both taper and parallel. They frequently have one *safe* side ; *i.e.* one side left smooth.

B. *Flat files,* amongst which a large proportion of the files used for mechanical work must be included, although many of them have special names by which their variations in width, thickness, and outline are distinguished. Such are :— *Hand files,* which are nearly parallel as to their width, but whose thickness is far from being uniform throughout. *Taper flat files* vary to a still greater extent in thickness, and also have their sides 'bellied.' *Parallel hand files* approach more nearly than hand files to uniformity of thickness and width, and *equalling files* attain to it in point of thickness, and are generally also parallel in width, and thinner than most of the preceding ones. *Pillar files* and *Cotter files* are of rectangular section, but narrower than hand files. *Mill-saw* or *Topping files,* both taper and parallel, are shown at *e* and *f* in Fig. 55. One or both of their edges are frequently rounded as in the half-section C, Fig. 56. Some hand, equalling, and other files have their edges similarly rounded, and some have bevelled edges, as in the half-section D in the same figure.

E. *Round files*, when small and taper, are known as *rat-tail files*. Under the name of *Gulletting files* (*c*, Fig. 55) parallel round files are very largely used for deepening the 'gullets' of the saw-teeth of that name.

F. *Frame-saw* or *Pit-saw files* (*d*, Fig. 55) are parallel (or sometimes taper) files of this section, which can be used for both 'gulletting' and 'topping.'

G. *Half-round files* are almost always taper. Those of the section F have more right to the name, but they are not included except as 'high-back half-round.' Most of the half-round files made in Sheffield are nearly of the proportions given in the section (G); but those made in Lancashire are of three different 'heights;' the thicker and the thinner sections being known as *high-back* and *flat-back* respectively.

H. *Cross* or *crossing files*, sometimes called *double half-round*. They are made both taper and parallel.

K. *Three-square files*, shown both taper and parallel in Fig. 55. They are very much used both for sharpening saws and for general purposes.

L. *Cant files*, generally parallel, and cut on all three sides.

M. *Knife-edge* or *Knife files*, either taper or parallel.

N. *Feather-edge files*, almost always parallel.

Many of the small files used by clock and watch makers, although the sections resemble those above given, are known amongst them by special names, derived for the most part from the particular use to which they are chiefly applied. Thus the crossing file (H, Fig. 56) in all probability owes its appellation to the convenience of its section for filing out the cross arms of clock wheels. Neither this nor the 'cant,' 'knife,' or 'feather-edge' files are often to be found in the workshop of the engineer.

Rubbers are large and coarse files, made of an inferior quality of steel, which are sold by weight. They are generally of the form and section given in Fig. 57, and are used for rough manufacturing purposes, but not for the higher kinds of work.

Bent files, called *Riflers*, are made of various degrees of curvature. One of them is represented in outline in Fig. 58. They are the only kind of file which can be brought to bear upon concave surfaces which have any considerable curvature in more than one direction, and they are therefore of occasional use in the workshop, especially among brass finishers, who use them for filing up concave mouldings in brass castings.

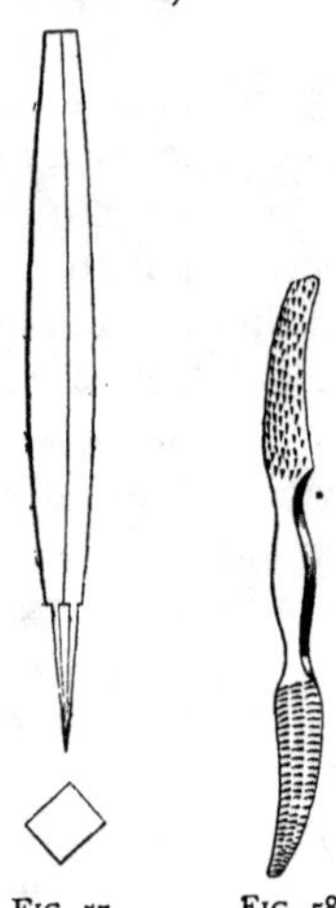

FIG. 57. Rubber. FIG. 58. Rifler.

With regard to the various kinds of *cut* by which file teeth are produced, we must mention that English files are divided into two distinct classes—*Sheffield files* and *Lancashire files* (Sheffield and Warrington being the places where they are chiefly made), and that these differ from each other in several particulars. Of these the most important formerly was the superiority of those made in Lancashire in respect both of form and durability, but, as far at least as the larger kinds are concerned, this no longer holds good to a sufficient extent to compensate for their higher price, although the small files used in clockmaking, &c., are still exclusively Lancashire. To one point of difference between them attention has already been called in connection with the 'half-round' section (G, Fig. 56); another is the fact of the 'cut' of Lancashire files, being only of the first kind given on the opposite page, so that our description must be understood to apply only to those sold as 'Sheffield.'

In Fig. 59 are represented the three principal kinds of *cut* by which files are divided into *double-cut* files (which are

DOUBLE-CUT.

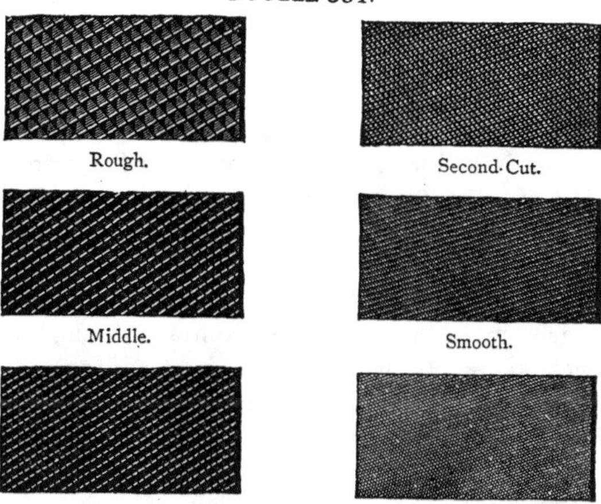

FLOAT-CUT.

Rough.

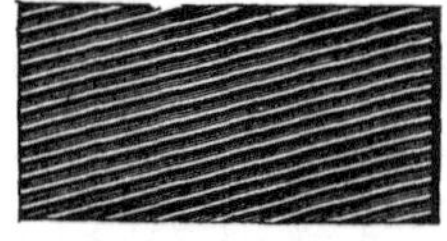

Bastard.

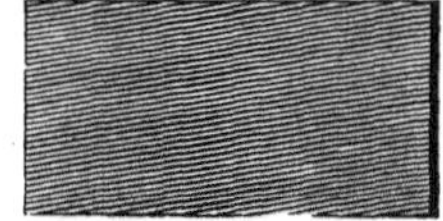

Smooth.

RASP-CUT.

Rough.

Bastard.

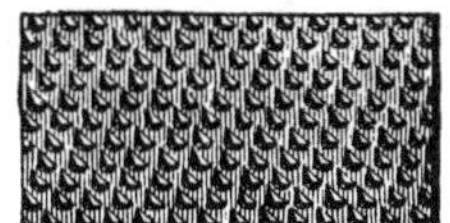

Smooth.

FIG. 59.

always meant when 'files' are spoken of without any particular cut being specified), *Single-cut* files or *Floats*, and *Rasps.* Besides these there is also another, which differs from the first only in the obliquity of its two series of cuts (one of them being almost square across the file): it is known as the *New cut*, and is said to have special merits for filing wrought iron. But of all of them the first is by far the most largely employed, almost all the files used for metal, as well as cabinet-makers' files, and many others, being double-cut. Saw files, however, are an exception, all varieties of them being frequently, though by no means invariably single-cut; these when in use being considered by many persons to 'cut sweeter' than those which are double-cut. To these the term 'float' is not applied; floats are extensively used for preparing brass for burnishing, rasps being chiefly used for wood, bone, and other comparatively soft materials.

In addition to the kinds of cut, the preceding figure shows also many of the degrees of fineness in which files are made, together with the names by which they are distinguished. The teeth are represented of the actual sizes which they would be in files 12 inches long, those of smaller files being finer, and those of larger ones somewhat coarser. The complete series is given in the case of the double-cut only, both floats and rasps being also made of about six different degrees.

As far as the double-cut is concerned, the engraving represents with tolerable truth the teeth of Lancashire as well as of Sheffield files, but among the former the finest are known as *superfine*, and not as 'dead-smooth.' The number and the accuracy of the teeth in some of these is certainly wonderful. Mr. Holtzapffel mentions having himself found that one of the smallest and finest Lancashire files contained nearly 300 cuts per inch of its length; yet these are produced entirely by hand—machine-cutting, although it has been carried out with some success on the continent of Europe, having hitherto made but little headway in this country. The operation is of such interest, as illustrating the very high degree of perfection which may result from skilful manipulation, that we cannot refrain from giving a slight outline of the process, which is conducted as follows :— The soft cast-steel *blank* is placed on an anvil with the tang towards the operator, and is there held by means of a foot-strap; a suitable piece of lead or pewter being interposed when its section requires a V-shaped or other special support, or when the under side has been previously cut. Commencing at the point, the file cutter then makes a series of cuts with a chisel of which the width rather exceeds the width of the cuts, striking it with a hammer which is strictly proportioned as to weight to the amount of burr to be thrown up at each cut, or in other words, to the degree of coarseness of the teeth. In order to place the chisel correctly he slides its edge a little way along the greased surface of the blank, till it comes into contact with the burr of the preceding tooth. The height of this burr, and the greater or less inclination given to the chisel, thus determine the space by which each succeeding tooth is in advance of the previous one. Double-cut files have two series of cuts inclined in opposite directions to the axis of the file, the edges of the teeth produced by the first series being very slightly smoothed over before the second series is commenced. For rasps the straight-edged chisel is replaced by a pointed punch, and in their case the evenness of the teeth depends solely upon the skill with which the file-cutter hops the punch into the right positions. In whatever pattern he may choose to arrange the teeth, his object should always be to place as few of them as possible exactly behind each other.

After being cut, files are straightened and hardened, during which process they are coated with a mixture which protects their teeth from the direct action of the fire.

Drills, which are, perhaps, the next most largely used of workshop appliances, trench so closely upon the province of 'machine-tools,' t at it is difficult to say how far they may be properly included in this chapter. For a drill without some mechanical arrangement by which sufficient speed or power for driving it can be obtained, is as useless as an auger without a handle, or a plane-iron without a stock. In default of a better distinction, we shall, therefore, only notice here some of the steel drills themselves, and the *portable braces* &c. in which they are sometimes used.

Most drills for metal have two similar cutting edges, a characteristic by which they may be broadly distinguished from those used for wood. This makes them, in a mechanical point of view, much more perfect instruments, as will be seen from the diagram Fig. 60, which gives sectional plans of a spoon-bitt (*a*), a centre-bitt (*b*), and a drill (*c*); the hole bored by each being represented by the circle. Supposing the rotary motion to be produced in each case by two equal and opposite forces, it will be noticed that in *a* one only of these forces is employed in cutting; the other being expended in merely pressing the back of the shell against the side of the hole, giving rise to the great friction and heating to which tools of this class are subject. In *b* a portion of this opposite force is utilised in driving the 'nicker,' the remainder producing a side strain upon the central pin. In *c*, however, both of them are wholly consumed in cutting, and the tool being in perfect equilibrium, it is possible for it to maintain the straightness of its course and the circularity of the hole without such support as is afforded by the shell in the one case and by the central pin in the other. But this distinction does not obtain universally, in each class there being some exceptions; such as the double spiral screw-auger previously mentioned (Fig. 41), and the Swiss drill figured on the opposite page, which are examples of double-edged wood borers and single-acting metal drills respectively.

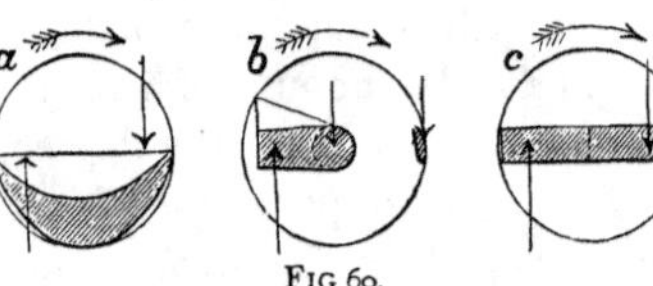

FIG 60.

In Fig. 61 the cutting portion of various kinds of drill are represented. Of these *a* is the ordinary watchmaker's drill, which although it scrapes rather than cuts, is more

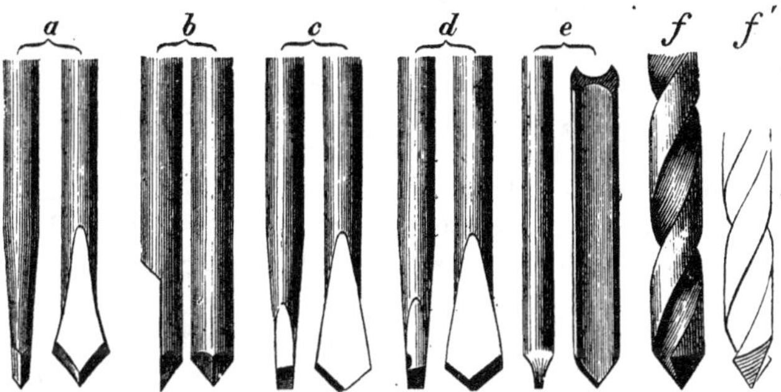

FIG. 61.—Drills.

efficient than might be expected, when its size is small and its revolution rapid. For a drill of this kind indeed, which is to be driven alternately backwards and forwards—which these generally are—double-bevelled edges are better adapted than those with a single bevel, since the latter, although they might cut more quickly when revolving in one direction, would do nothing when moving in the other. One great disadvantage in this form of point is that the diameter of the hole produced by it is altered every time the drill is sharpened; which, however, can be obviated by making its sides parallel for a short distance above the cutting edges. From doing this another result—and a most important one—also follows; viz., the sides are enabled to assist in the guidance of the drill, which is thus rendered much less liable to change its direction during its advance, in consequence of any slight alteration in the direction of the pressure applied to it, or any other cause. With a drill of the form *b*, which also cuts when revolving in either direction, it is almost impossible to drill a hole otherwise than straight; and however frequently it may be sharpened its original diameter is maintained. Of the drills which are adapted for revolution in one direction only, *c* is more frequently to be met with than any other; its use being general, alike for the smallest perforations produced in a foot-lathe, and for the heaviest operations of a drilling machine. It may be observed that these have but little inclination of their own to enter the material which is being drilled, since the angle of their cutting edge is by no means acute, and its front face is either at right angles to the surface upon which it is operating, or is slightly tilted forwards. This, combined with the bluntness of the point—which is an evil inseparable from this form of drill—necessitates, in the case of the harder metals at least, the application of considerable *feeding* pressure that a drill of this kind may be kept up to its work; its action is then not otherwise than rapid, and is perfectly analogous to that of the metal-cutting tools used in the lathe. The great heat evolved, however, forbids any but a very moderate rapidity being given to it. By simply forming a groove in the front face of each of the cutting edges, as shown in that marked *d* in the figure, any desired degree of acuteness can be given, the strength of edge required for the particular material under treatment being the only limit to the extent to which this may be done. To this plan, however, the much greater amount of grinding necessary whenever the edges of the drill require to be renewed is an objection.

Two kinds of drill, in which the principle of making them self-guiding by providing them with parallel edges is fully carried out, are shown in *e* and *f*. The former is known as the *fluted* drill, and the latter is the *twist* drill. Their difference, as far as their cutting edges are concerned, corresponds with that just pointed out with reference to the drills *c* and *d*. In *e* the front face of each edge (which is formed by one side of the groove or fluting) is perpendicular to the surface of the work, so that the depth of its cut is dependent upon the feeding pressure. But in *f*, by the simple and beautiful expedient of carrying the groove spirally round the stem, any moderate inclination can be given to this face, and however much the extremity may be ground away, its angle always remains invariable. When the twist is as sharp as that shown in the figure, this drill is admirably adapted for boring the hardest kinds of wood or the softer metals, which it cuts greedily with but little pressure. For iron and similar materials, its strength—which is inferior to that of *e*—is scarcely sufficient. In sharpening either of these drills, it is necessary to grind the extremity in such a manner that the lower face of each cutting edge shall form a portion of an independent spiral, instead of allowing both of them to become part of one and the same conical surface. In *f* the spiral face is shown much exaggerated, but its pitch may be slightly varied according to the nature of the material to be drilled and the extent to which the strength of the edge may be reduced in operating upon it. But however strong an edge may be required, the point must never be absolutely conical, of which the effect would be similar to that of grinding the drill *c* without any bevel (*i.e.* with the cutting angle equal to the following angle of its edge), in which case no available pressure would suffice to feed it. Consequently the small angle which this face forms with the surface of the work—which corresponds with that called by Mr. Babbage the 'angle of relief' in the case of metal turning and planing tools —is absolutely essential to the working of a drill; and with every increase in this angle a correspondingly diminished pressure is capable of giving an equal amount of feed; in other words, the more greedy the drill becomes.

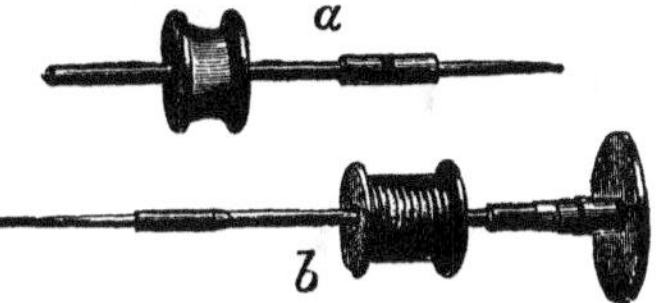

FIG. 62.—Bow-Drill Stocks (¼).

In drilling holes of very small diameter, we have seen that the pointed form of tool employed by watchmakers and others requires rapid rotation alternately in opposite directions, with but little feeding pressure. These conditions are well satisfied by the ordinary bow-drill (*a*, Fig. 62), the method of using which will be evident from Fig. 63. The

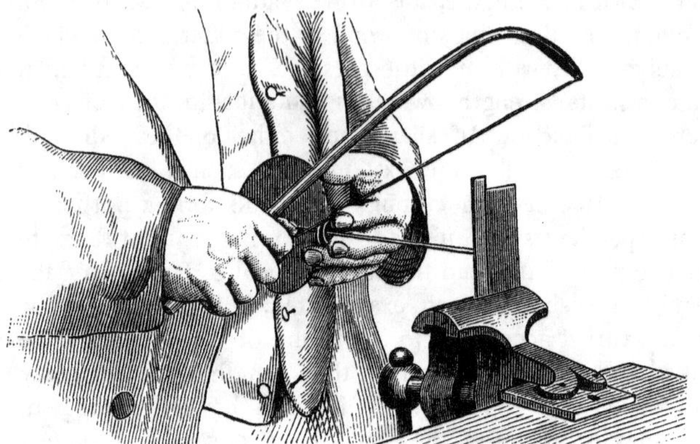

FIG. 63.—Drilling with Bow-Drill.

drill is either fixed in a stock, as shown above, or is itself of sufficient length to carry a wooden sheave or barrel, round which, when in use, the string of the *drill-bow* takes a single turn, the pointed end of its stem resting meanwhile in a shallow centre hole in a sheet-iron *breast-plate*, strapped round the chest of the workman ; or, if the work cannot con-

FIG. 64.—Archimedean Drill-Stock (¼).

veniently be so fixed that the drill may be held horizontally (or nearly so), the breast-plate may be held in the left hand, by which, when the drill is small, ample pressure can be applied.

In *b*, Fig. 62, is another form of bow-drill stock, for which no loose breast-plate is required. A coarse screw-thread cut upon the surface of its barrel saves the gut, of which the bow string is generally made, from chafing at its point of crossing. The bows in question, when small, are almost invariably made of whalebone, larger ones being generally made of cane, but occasionally of steel.

The *Archimedean drill-stock* (Fig. 64) is an invention which dispenses with the necessity for either breast-plate or bow, the motion of the right hand being, by means of the spirally grooved stem, converted into rotation, which in its constant change of direction resembles that obtained from the drill bow. But it is by no means an equally economical application of power, and—probably on that account—it has not at present done much towards supplanting the bow-drill amongst those who have much use for such instruments. The many-threaded spiral shown in the engraving is produced by twisting a piece of pinion wire—but the grooves of these drill-stocks are frequently much fewer in number and of larger size.

The *bevel-wheel drill-stock* or *brace* (Fig. 65) may in many cases be conveniently used for drilling holes which are of larger size than those for which the foregoing stocks are adapted; but which, at the same time, are not so large as to require a greater amount of feeding power than is obtainable by letting the flattened head bear against the chest of the workman. Its construction is sufficiently clear from the engraving; but the proportion between the bevel wheels in some of these stocks is such as to give a greater, and in others to give a less velocity to the drill than that of the winch handle by which it is driven.

Passing on to yet larger sizes, for which the above method of feeding the drills is inadequate, we come to the old—but simple and efficient—*smith's brace*, which is represented in position for use in Fig. 66. It will be observed that it requires to be constantly associated with a *cramp*, one end of which must be provided with a feeding screw; unless indeed the brace itself contain one, in which case the cramp may consist simply of a stout bar bent to the requisite form. (In the engraving both the cramp and the brace have feeding screws, but this is superfluous.) For drilling small articles, the cramp must in turn be supported in a *tail-* or

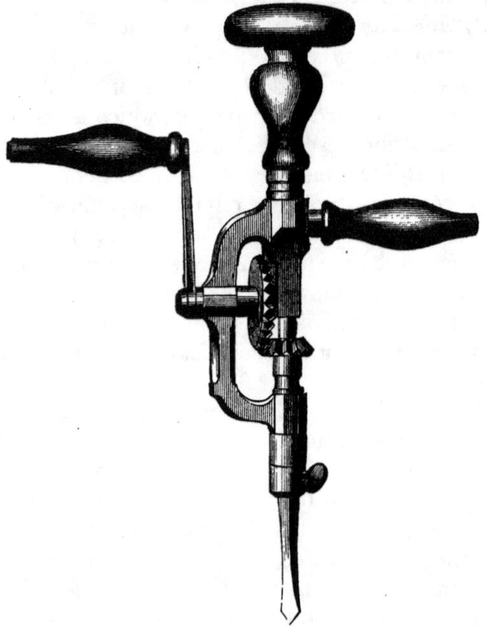

FIG. 65.—Bevel-wheel Drill-Stock ($\frac{1}{4}$).

bench-vice (as in the figure), or the place of this may be taken by a *pillar*, either temporarily or permanently fixed to the bench, on the surface of which the work is then placed, an adjustable horizontal arm carried by the pillar receiving the thrust from the upper end of the brace. Another method of feeding the drill, which has long been employed in blacksmiths' shops, is to arrange a lever in such a manner that a heavy weight at one of its extremities may constantly press downwards upon the brace placed below it near the other; but this, although it is superior to the cramp or pillar arrangement, since it gives a continuous and self-acting feed, can hardly be classed amongst portable forms of drilling apparatus.

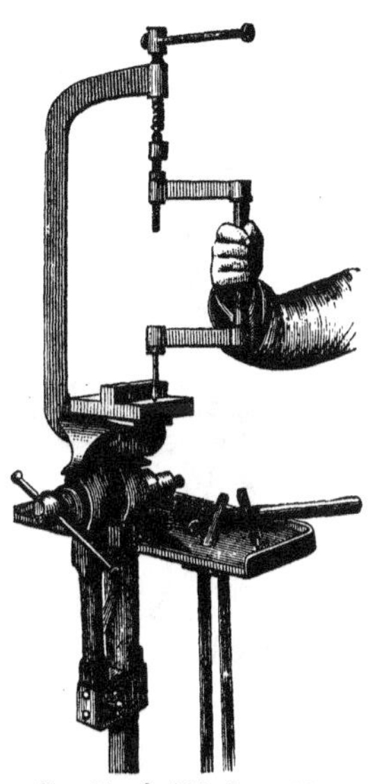

FIG. 66.—Smith's Brace (⅛).

When portability is a main consideration — more especially as in such cases the space required for working an ordinary brace is frequently not available —some kind of *ratchet-brace* is almost always substituted for it. This admirable contrivance is represented in Fig. 67 almost in its original form, and although much ingenuity has been expended in devising modifications of it, few, if any of them, have obtained as general acceptance. The stock, in the lower part of which the drill is fixed, contains its own feeding-screw, of which a portion appears in the engraving below the elongated hexagonal nut. By this nut, which forms the entire upper portion of the stock, the feeding screw is wholly concealed when it is screwed home. The hand-lever on the right in the figure is entirely detached from the body of the stock, but by means of the ratchet-wheel on the latter and the click carried by the lever, the effect of imparting to it any amount of backward and forward motion is to cause an intermittent revolution of the stock together with any drill which may be fixed in it. At the same time any desired feeding pressure can be obtained by simply arresting the upper portion of the stock during the forward movement of the lever. A short length of the feeding screw is thereby withdrawn, so that the distance between the point of the drill and the top of the stock is increased. A simple cramp is of course in general a necessary adjunct for this as well as for the plain brace, but its height can be very much re-

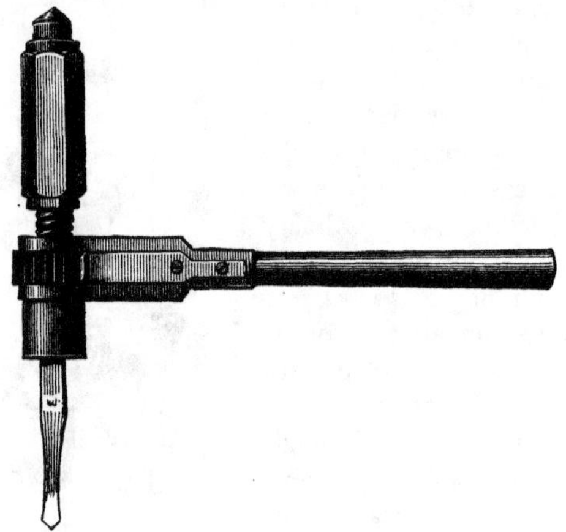

FIG. 67.—Ratchet-Brace ($\frac{1}{4}$).

duced; as can also the distance from any high projecting portion of the work at which the drill can be applied with this instrument.

After a hole has been drilled, its sides not unfrequently require smoothing, or it may have to be enlarged. For either of these purposes *rimers* or *broaches* are the tools generally used; their forms and their cross-sections, as also their sizes, varying according to the purpose for which they are intended. Two 'taper' rimers are shown in Fig. 68, in which the angles do not exceed 90°, and are favourably placed for cutting rather than for scraping, which is not the case when the cross-section consists merely of a flat-sided figure with five, six, or even eight sides, as it sometimes does. In using rimers of this kind, a very trifling downward pres-

required, for their finely tapered form causes this to iverted into a powerful lateral thrust against the sides e hole, which in their case constitutes the feeding pressure. For leaving the sides of ole parallel instead of making it taper, ers are not unfrequently made of equal eter for the greater part of their length, the le of the cutting being then performed by sharply tapered extremity with which they rovided, the upper portion serving only as le. Or a similar effect may be produced ing them a 'bellied' form and retaining the g angles throughout their whole length, but these the perfect straightness of a hole nnot be secured with equal certainty.

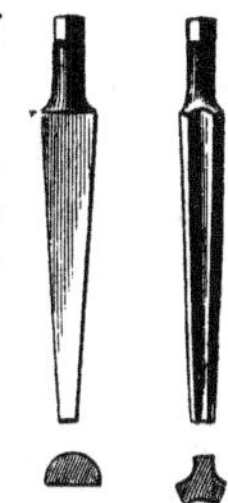

FIG. 68. Rimers.

a rimer be made, not from a plain conical piece of steel, from one on whose surface a screw-thread has been viously cut, it is evident that if its advance at each olution be equal to the pitch of its screw, it will produce unterpart of the thread in the interior of the hole ugh which it is passed. Exactly on this principle are *screw-taps* by which internal or 'female' screws are pro-ed, to which, together with the usual means of cutting ernal screws by hand, most of our remaining space must be devoted.

Internal screws, then, were formerly made by applying to the hole which was to be 'tapped' a tapered screw-tap of hardened steel, upon the conical surface of which a screw-thread had been previously cut to an equal depth throughout its length, three or four flat facets having been afterwards filed upon it to at least the depth of the thread. Its section thus resembled *a* or *b* in the accompanying cut (Fig. 69), from which it will be evident that—especially in the case of *a*—the exceedingly obtuse angles which they

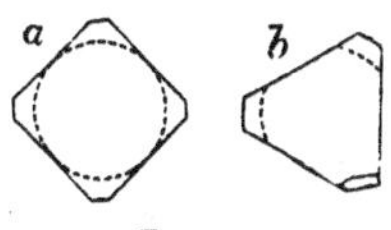

FIG. 69.

presented were quite incapable of cutting, only remove the material from the hollow port thread by a slow scraping action. This, toget considerable amount of burring or pressing up of the parts of the thread, enable them to operate with efficiency upon wrought iron and other malle but with cast iron they were almost entirely incom deal. Under the most favourable circumstances the was a very slow one, the tap requiring to be wor wards and forwards several times as each slight adv made; the full depth of the thread in a nut or in 'thoroughfare' hole being attained only when the the screwed portion of the tap had passed through siderable force was necessary to cause the tap to a all, and this was applied by means of a long ba *wrench* with a square hole at its centre, into which squared upper end of the tap was fitted. But in o give it sufficient strength, this square portion wa larger than the rest of the tap; consequently, after work, one of these old taps could not be removed hole without being unscrewed through its whole which entailed much waste of time.

Many improvements have been introduced into th 'screwing tackle' used at the present day, to the tion and manufacture of which Sir J. Whitworth has much attention. Fig. 70 shows a set of three screw-taps, consisting of a *taper*, an *intermediate*, and a *plug tap*. A cross section to a larger scale—which applies equally to all of them—is also given in the figure; by which the great reduction in the cutting angles and the superiority of their position with regard to the surface to be cut, which results from the adoption of this fluted form, is rendered clear. For in the sections *b* and *a*, Fig. 69, these angles exceeded 120° and 135° respectively, whereas in the present case they are not more than 90°, and the front face of each (the arrows showing the direction in which the tap revolves when work-

sure is ...l, so that it can remove the metal from between be cor... by a true cutting action, and has no tendency to of th... burr in the manner true ...ed above. The threads a h... taps themselves are rin... parallel through-dian... pered form of the in-who... e and taper taps being the ... ubsequently taking are p... ps of the threads to a guid... or less extent. In by gi... r this is only done cutti... t distance from the with ... remity, but in the car... carried to such an ...t that the threads are en-b... obliterated at the bottom pre... n, a small number of rev... nly at the upper end a ... their full depth. This thr... a taper tap to enter the ... of the proper size du... difficulty, its first few revolutions doing little more ext... e out a thread of sufficient depth to lead it on- ... But by the series of cutting angles which it presents during its advance—each of which to a small extent exceed its predecessor in height—successive portions are remov... from the spiral groove through which they pass, till on its upper end emerging from the hole, the tap will be found have left behind it an almost perfect counterpart of ...thr...d which was originally cut upon its own surface. ... inequality which it may possess is then easily re-mov... by following it up with the intermediate tap; the tedious process of unscrewing mentioned above being now unnecessary, since the more perfect action of these taps enables the force applied to ... to be reduced, so that the

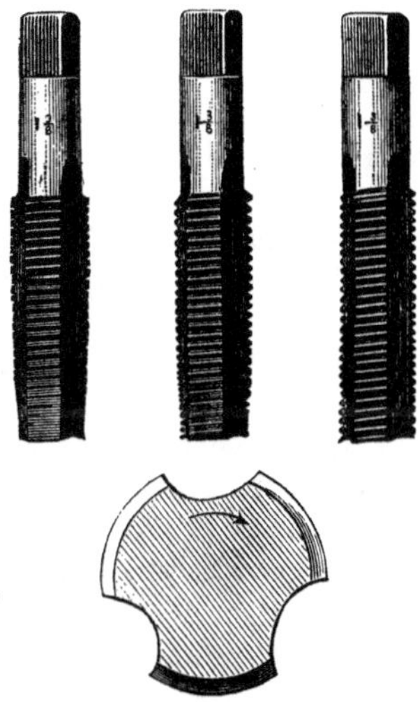

FIG. 70.—Set of Screw-Taps, with enlarged section.

square head can be made sufficiently small to drop through the hole through which the screwed portion has passed. If the hole be shallow, the operation must be commenced with the intermediate instead of with the taper-tap; in this case the finishing cut is given with the plug-tap, of which the threads are left of the full depth quite to the extremity. It should be noticed with reference both to taps and rimers of the above, or any other section with several grooves or flutings—unless, indeed, the proportion of the circumference occupied by the grooves is greatly in excess of what remains—that it is essential to their efficiency for their advancing or cutting angles to project beyond or to be at a greater distance from the centre of the figure than the following angles, which are not required for cutting. If this be not attended to, it is evident that the whole of the concentric portion of the section must be forced into the material before any feed can be obtained; which, even if it were possible, would cause so great an amount of friction against the interior of the hole, that the tool would be broken by torsion before it could be made to revolve. On carefully observing any well-made taps, like those shown in the engraving, it will be seen that on this account the threads, more especially in taper taps, by which the bulk of the material is removed, are always filed off to a much greater extent upon their following than on their cutting angles. Like the rimers of similar section noticed above, they can therefore only cut whilst revolving in one direction. In the old flat-sided taps this was not done, and in spite of the circumference being cut away to an extent which greatly endangered the evenness of the thread produced by them, the necessity for forcing these still comparatively large surfaces into the material before any feed could be obtained, had doubtless more to do with their enormous and unnecessary consumption of power than even the unfavourable nature of their angles. This view is confirmed by the much greater ease and efficiency with which six-sided and eight-sided rimers perform their work;

for in these, since no part of the conical surface remains, a very much smaller amount of pressure suffices to cause the angles in which their facets meet to enter the work.

In cutting external screw-threads also much improvement has been effected since the days in which the plain *screw-plate* was the only hand-tool used for the purpose. It consisted simply of a steel plate with a number of tapped holes in it, a series of from two to six of which were to be used for each size of screw. These were even less capable of cutting than the early forms of tap just described, the screws being entirely produced by pressing or swelling out of the threads at the expense of the material between them. The diameter of the finished screw was thus generally greater than that of the original wire or rod. By the addition of two or more notches to each hole a certain amount of cutting took the place of this violent treatment; and thus modified the *notched screw-plate* (Fig. 71) is still much used for small work.

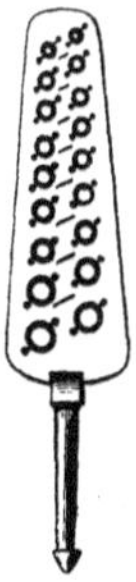

FIG. 71. Screw Plate.

But for larger screws it became absolutely necessary to make the cutting action more perfect, and so to diminish the wasteful expenditure of power which the use of the screw-plate entails. This was done by the introduction of the *stock and dies* (Fig. 72), which enabled one pair of dies first to trace out the thread lightly upon a cylindrical surface, and eventu-

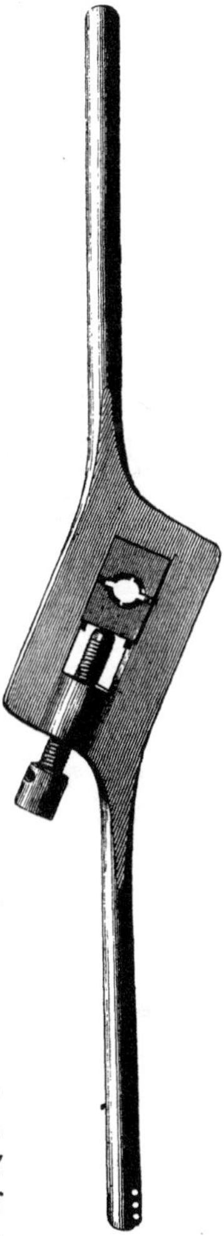

FIG. 72.—Ordinary Stock and Die (¼).

ally, by being gradually brought closer together, to cut it to the full depth. Moreover, by making the cavity in the die-stock of sufficient length (as in the engraving), or by various other means, the dies could be easily removed and replaced by others of different sizes. This arrangement, however, is defective, inasmuch as the curvature of the dies must be either too great for them to admit the uncut bar—in which case their outer angles only can act,

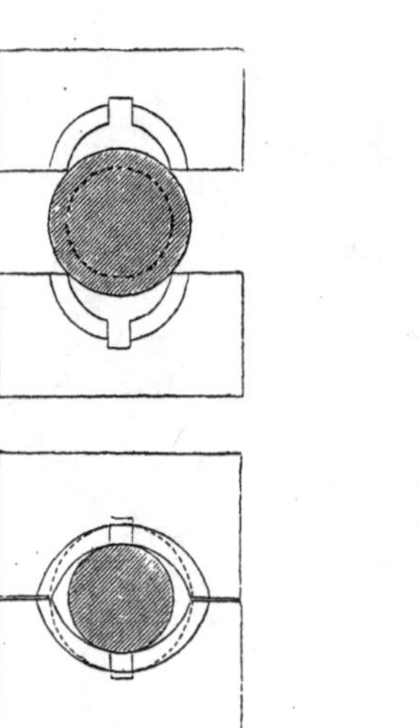

FIG. 73.

FIG. 74.—Whitworth's Stock and Dies.

which, having but little guiding power, are liable to set out a thread which is uneven or 'drunk;'—or, as the screw approaches completion, the work is performed more and more by the central portion only of each die. Fig. 73 shows these defects, the curvature of the dies coinciding with that of the finished screw in the upper half of the diagram, and with that of the uncut rod in the lower half. The usual practice is, therefore, to cut the dies with a curvature intermediate between the two, but this is only a partial remedy for the evil.

A much more satisfactory arrangement is that originally devised by Sir J. Whitworth, in which three dies of small but not of equal width are substituted for the two wide ones, shown in the foregoing figures. As we have seen in the case of screw-taps, the effect of narrowing the concentric portions is to effect a corresponding reduction in the power which must be expended in working them; the necessity for retaining sufficient width to ensure the evenness of the screw-thread being the chief limit to the extent to which this may be done. In the *Guide screwing-stock*—of which the central portion is represented in Fig. 74—the advantages of efficient guidance combined with great facility of working are obtained by making two of the dies only of small width, and performing the bulk of the cutting with them; employing the third almost exclusively as a guide to ensure the uniformity of the thread. This—the stationary die (marked *a*)—is clearly seen in the engraving, as are also the moving dies, *b* and *c*; the top plate—which to a great extent conceals them when in use, and of which the holding down screws appear—having been removed for the purpose of showing them. These moving dies are made to advance equally by turning the nut (*e*), by which the inclined portions of the sliding-piece (*d*) are gradually drawn in behind them till the full depth of the thread has been attained. Their action, however, is not simultaneous, the curvature of their extremities and their cutting angles being so arranged that one shall cut when the stock is turned in one direction, and the other when it is turned in the opposite one. These dies can be replaced by others for the production of screws of other diameters, and can be sharpened by grinding with the same facility as ordinary tools.

In cutting the threads upon the dies themselves, *Master-taps* are used; one of which is shown in Fig. 75. These differ from the ordinary taps shown on a previous page (see Fig. 70), in being of larger diameter (Sir J. Whitworth's exceed the working taps by twice the depth of the thread, which gives perfect contact, and thus produces an even thread at the commencement of the cut), and also in having much narrower longitudinal grooves and a larger number of them, which is rendered necessary by the small amount of surface which the dies present. Similar instruments, known as *hobs*, are also employed in forming the cutting ends of screw-chasing tools for use in the lathe. The importance of possessing and carefully preserving a correct set of master-taps will be evident when it is remembered that the screws produced by the above processes are in all cases mere copies of those already in existence upon the surface of the taps and dies; and that the simplest, if not the only practically available method of maintaining the uniformity of these threads—which it is in the highest degree desirable to do—is to have at hand a constant standard of reference by which the effect of continued wear upon the working copies may be detected and remedied.

FIG 75. Master-Tap.

So important does Sir J. Whitworth consider the subject of the uniformity of screw-threads—not only of those made in a single workshop or for one particular purpose, but of all to which a fixed pitch can be applied throughout the whole country—that he arranged and published partly in the year 1841 and partly in 1857 the following table, the use of which, for machines and engineering work, has been gradually extending with most beneficial results; the excessive loss and inconvenience which arose from the absence of any standard pitch being now universally acknowledged. (See table next page).

The angle made by the opposite sides of the thread is 55° in every case, but the extreme depth which this angle would give is reduced by rounding off the top and bottom,

Table of Pitches for Screws with Angular Threads.

No. of Screw Threads per Inch	Old Sizes	New Standards of Size. Decimals of an inch	No. of Screw Threads per Inch	Old Sizes	New Standards of Size. Decimals of an inch	No. of Screw Threads per Inch	Old Sizes	New Standards of Size. Decimals of an inch
48		·100	12		·525	4½	2⅛	2·125
40	⅛	·125	12		·550	4	2¼	2·250
32		·150	12		·575	4	2⅜	2·375
24		·175	12		·600	4	2½	2·500
24		·200	11	⅝	·625	4	2⅝	2·625
24		·225	11		·650	3½	2¾	2·750
20	¼	·250	11		·675	3½	2⅞	2·875
20		·275	11		·700	3½	3	3·000
18		·300	10	¾	·750	3¼	3¼	3·250
18		·325	10		·800	3¼	3½	3·500
18		·350	9	⅞	·875	3	3¾	3·750
16	⅜	·375	9		·900	3	4	4·000
16		·400	8	1	1·000	2⅞	4¼	4·250
14		·425	7	1⅛	1·125	2⅞	4½	4·500
14		·450	7	1¼	1·250	2¾	4¾	4·750
14		·475	6	1⅜	1·375	2¾	5	5·000
12	½	·500	6	1½	1·500	2⅝	5¼	5·250
			5	1⅝	1·625	2⅝	5½	5·500
			5	1¾	1·750	2½	5¾	5·750
			4½	1⅞	1·875	2½	6	6·000
			4½	2	2·000			

each to the extent of one-sixth, as the diagram (Fig. 76) will explain. Thus the depth given to the thread is only two-thirds of that which it would have if its sides intersected, being ·64 of the pitch instead of ·96.

Fig. 76.

Square-threaded screws, which are used for many purposes, since they possess greater *power*, although less strength, than those of the above form (as explained in another volume of this series),[1] have generally half the number of threads per inch than would be given to them if the threads were angular, the depth being usually equal to the space between the threads. When coarse, however, they should be deeper in proportion than when they are fine.

[1] *Elements of Mechanism*, p. 17.

For many purposes—such as screws cut on the outside or inside of tubes—much finer threads than those given in the preceding table are required. In these there is much less uniformity than could be wished, although the immense consumption of metal tubes for the distribution of coal-gas has caused some such threads to be pretty generally recognised, under the name of *gas-threads.* The following are the pitches used by Sir J. Whitworth in his taps and dies for gas-tubing :—

Diameter of tube, in inches	$\frac{1}{8}$	$\frac{1}{4}$	$\frac{3}{8}$	$\frac{1}{2}$	$\frac{3}{4}$	1	$1\frac{1}{4}$	$1\frac{1}{2}$	$1\frac{3}{4}$	2
Number of threads per inch	28	19	19	14	14	11	11	11	11	11

This and the previous table must not, however, be supposed to apply only to the forms of apparatus which we have been describing—for this would confine their use within very narrow limits, since but a small proportion of the screws

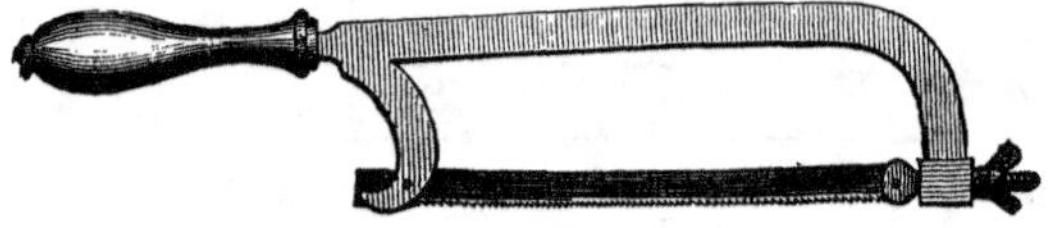

FIG. 77.—Smith's Saw ($\frac{1}{6}$).

in daily use are cut by hand with their assistance. For the production of many kinds, indeed, stocks and dies, even of the best construction, are much less well adapted than are other instruments which we have occasion to describe hereafter; amongst which may be mentioned hand-chasing tools, the screw-cutting lathe, and bolt-screwing machinery, all of which have their special advantages.

The remaining hand-tools for metal, of which the use is sufficiently general to demand our attention—if at least we except the Smith's saw (Fig. 77), which is almost the only representative of its class—will be found to consist chiefly of those which effect their operations by some kind of punching or shearing, rather than by what we have been considering as a true cutting action. In the latter, the chippings, filings, &c., produced, are almost invariably so much waste material; in the former, the portion removed is abstracted in a single piece, and frequently constitutes the more valuable part of the product. Another important difference distinguishes *punches* and *shears* from the tools with which we have been dealing, viz., that in their case no feeding pressure is required, so that the direction of the driving force applied to them determines that in which their action takes place.

The term 'punch' is made to include two very different kinds of instrument. Of one of these the chief duty is to indent the material to a greater or less extent, without absolutely dividing it;—examples of which are to be found in the pointed centre-punch, the various thick-edged circular and other punches with which thin discs, &c., may be partially cut from sheets of metal, and also in the ordinary cold-chisel, when used for nicking sheets or bars of moderate substance. But these cannot be driven through the entire thickness of the material, and therefore cannot be considered to cut it, unless it be placed upon some hard and ponderous mass, such as an anvil, by the reaction of which they are then assisted. In the accompanying diagram (Fig. 78), *a* represents in section a tool of this description partially driven through a sheet of metal which rests upon the hard surface of an anvil. The difference between this and a punch of the second kind (*b*)—which must be used in conjunction with a *bolster* placed beneath the work, and is thus precisely analogous in its action to the pair of shear blades (*c*)—is at once apparent; since in these cases the portion of the tool which is below the work is almost exactly similar in figure to that which is above it, and, except when pre-

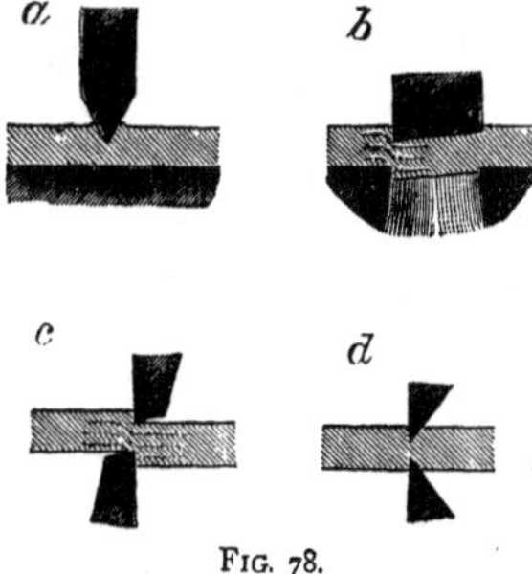

FIG. 78.

vented by extraneous circumstances, such as inequality in their areas, each performs an equal share of the work. It is evident, however, from the obtuseness of the edges in both cases, that the division of the material is effected by tearing its fibres apart rather than by cutting them fairly asunder; and this is equally true of the upper and lower blades of a pair of shears, and of the punch and bolster, which may be regarded as similar blades of a circular or other figure.

Two pairs of *hand shears* are represented in Fig. 79. The upper one of them is the ordinary form of *English hand shears*, which are often called snips, to distinguish them from the fixed-bench shears used for metal of greater thickness than can be cut when the tool is merely grasped in the hand, their blades being occasionally bent instead of

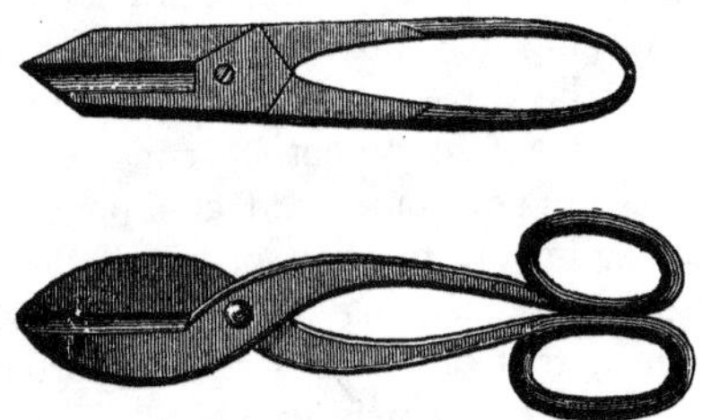

FIG. 79.—Hand Shears.

straight, for greater convenience in cutting out curved work. The lower pair, provided with bows like scissors, is of that known as the *Scotch* pattern.

For cutting wires, &c., a modified form of shearing tool is sometimes employed. Opposite the pivot in each half of a pair of single-jointed nippers a notch is cut. These notches coincide with one another when the nippers are open, so that a wire can be inserted into them. On closing the nippers, the wire is easily and cleanly divided. But these tools are but little used, either the *cutting nippers* or the *cutting pliers* shown in Fig. 80 being ordinarily employed for the purpose. In these, although the direction of their edges is different, the cutters are alike, the section *d* (Fig. 78) applying equally to both of them. From this section it will be perceived that their action differs entirely from shearing, their edges, although perfectly similar, being merely two opposed wedges which can be brought together till they meet, but which cannot pass each other, after the manner of shear-blades; consequently their angles are very much more acute, being from 30° to 40° instead of from 80° to 90°, to which the edges of shear-blades for metal are generally ground.

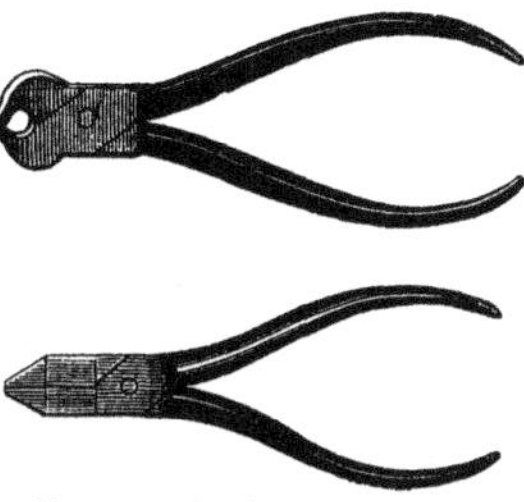

FIG. 80.—Cutting Nippers and Pliers (¼).

Of the use of punches combined with bolsters—which we have been obliged to introduce above—it is hardly possible to give an example without trespassing into the province of machine-tools. For their performance clearly depends upon exact correspondence being obtained between the position of the punch on the upper side of a plate and that of the bolster (or *die* as it is called when applied to the larger forms of punching machine) on the under side. Various kinds of holder, by the upper part of which the punch shall be guided when driven, whilst the lower either receives or itself forms the bolster, may be and possibly are employed; but some form of machine, in which both the punch and die can be fixed, and powerful pressure can be applied, add so greatly to their value that they are almost invariably so used

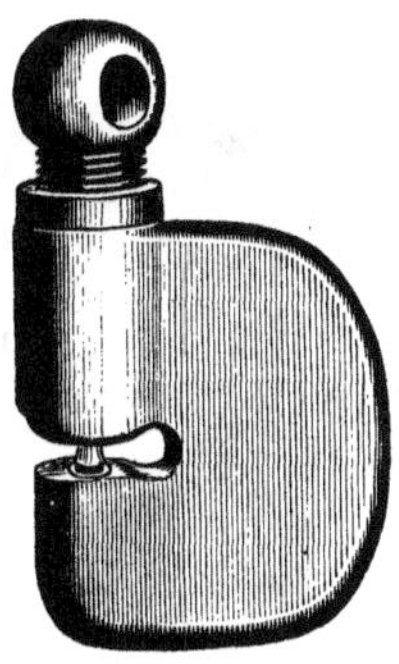

FIG. 81.—Punching Bear.

in engineering workshops. The simplest and most portable of these hand-machines is the *punching bear* (Fig. 81), which is, in fact, a powerful form of screw-press—by which plates of considerable thickness can be perforated by hand power. Applying his strength at the extremity of a lever passed through the eye of the screw, one man is able with this instrument to punch a hole as much as $\frac{3}{4}$ of an inch in diameter through a plate $\frac{5}{8}$ of an inch in thickness; and other equally powerful forms of apparatus are made.

Yet one other metal-tool must be mentioned here, which although it may be sought for in vain in the most complete tool-manufacturer's list, gives invaluable assistance to the accurate worker in metal. We allude to the *scraper*, the adoption of which for the production of plane metallic surfaces in lieu of the objectionable practice of grinding them, has resulted in a very much nearer approach to perfection than was previously possible. To this subject we shall have to return in the next chapter: at present we confine ourselves to the tool itself; which may most conveniently be made by grinding the end of a worn-out three-square file to the shape shown in Fig. 82. The three sharp edges thus produced must then be carefully set on an oilstone, the operation being frequently repeated when the tool is in use.

FIG. 82.—Metal Scraper.

Although the grinding of metallic surfaces with abrasive substances—such as *emery*—is by no means to be recommended, cases may sometimes occur in which the extreme hardness of the material or other causes render this treatment necessary. The following are some of the usual methods of performing such an operation. Either two surfaces may be simultaneously treated, the emery—reduced to a powder of as even a grain as possible, which can best be effected by 'washing' it—being placed between them with a small quantity of oil; or the powder may be applied to the surface of a metallic wheel or *lap*, which is then made to revolve, and is used in the same manner as a small grindstone; or it can be glued upon sheets of paper and calico, which are thus converted into *emery-paper* and *emery-cloth*. These last are the forms in which emery is most frequently used in the workshop, pieces of them being wrapped round a file or strip of wood, and applied to the surface of the work, either with or without oil, which leaves the work dull, though it causes the emery to cut more smoothly. For some purposes, again, *emery sticks* are preferred, the powder being in their case attached directly to the surface of the wood; whilst for others it is more conveniently applied in the consolidated form of *emery* or *corundum-wheels*. The marks by which the different degrees of coarseness of emery-cloth are ordinarily distinguished in London, are:—No. 0, No. FF, No. F, No. 1, No. 1½, No. 2, No. 2½, and No. 3; the first of these being the finest, and the last the coarsest. Emery powder itself is sold under the following names, commencing with the coarsest, of which the grain is about the size of mustard seed:—

Corn emery.
Coarse grinding emery.
Grinding emery.
Fine grinding emery.
Super-grinding emery.
Coarse flour emery.
Flour emery.
Fine flour emery.
Superfine flour emery.

The other abrasive materials used in the treatment of metallic surfaces are not often to be met with in engineering workshops. They are chiefly confined to the arts, in which smoothness of surface or 'finish' is a more important consideration than accuracy of form, and they are less frequently applied to wrought or cast iron than to brass and similar metals. Strips of oilstone, or of Water of Ayr stone (which is best kept constantly damp, so as to prevent its becoming hard, and should be used with water) may occasionally indeed be employed for removing the marks of the file or the scraper; but for working surfaces—upon which the chief care should always be bestowed—this treatment is neither necessary nor desirable. One rather favourite method of finishing flat portions of the exterior of a piece of work by 'curling' the surface was introduced by the elder Mr. Holtzapffel. The account of the process given by his son is as follows :—

'The work,' after being filed, scraped, and stoned (with Water of Ayr stone), is 'clouded with a piece of charcoal and water, by means of which the entire surface is covered with large curly marks, which form the ground. The curls resemble an irregular cycloidal pattern, with loops of from $\frac{1}{4}$ to 1 inch diameter, according to the magnitude of the work. Similar but smaller marks are then made with a piece of snake stone, blue stone, or even a common slate pencil filed to a blunt point. The general effect of the work much depends upon the entire surface being uniformly covered; with which view the curls should be first continued around the margin; the central parts are then regularly filled in; after which the work is ready to be varnished.'

In taking our leave of the subject of the hand-tools used for metal—from which we have been rather digressing—we must not, however, omit one important subject in connection with them; namely, the maintenance of their cutting edges, which are for the most part liable to speedy deterioration, owing to the hardness of the materials upon which they are used. In one large class indeed which we have been considering—the files—the acuteness of the edges, when once lost, cannot be recovered; in others, treatment upon the grindstone, as in the case of wood-tools, suffices for their renewal; but for the most part the best, and frequently the only available method of restoring their lost powers of cutting is to have recourse before grinding to the processes of forging and tempering. Into the consideration of the former of these indeed we cannot enter; the subject is far too wide to be treated in the cursory manner which would be absolutely necessary here, and, moreover, a few practical lessons in the art are worth more than a volume of description; but with regard to the latter we cannot but think that a few pages may be profitably devoted to it. For although the number of tools mentioned in the present chapter which require this treatment at the hands of the user as well as of the manufacturer is not large, it must be remembered that the cutters of almost all machine-tools also are maintained in a state of efficiency by similar means.

In the first place let us clear up one point in connection with the word *temper*, which may otherwise lead to confusion. As used by the steel manufacturers—on whom so very much depends in the matter of cutting tools—this term is almost synonymous with the *degree of carbonization* of any particular sample of steel; that which is highly carbonized, or contains a larger proportion of combined carbon, being considered to be of a 'high temper,' and that in which the proportion is smaller, of a 'low temper.' Thus, the steel which, during cementation, has received most carbon, will be said by the manufacturer to be of the 'highest temper,' and therefore, suitable for tools which are to be used for turning or boring cast steel; the next lower being adapted for ordinary turning tools, the next again for chipping chisels, and that of the lowest temper for taps and dies, or for the knives of shearing machines. So it will be seen that it is always desirable to state the particular purpose for which the steel is to be used, and therefore, to make a practice of doing this, and then to leave the selection to some respectable manufacturer, is, we believe, the best recipe we can give for obtaining good steel.

Crucible cast steel of the best quality should always be employed for metal-tools, for its greater cost in the first instance is more than repaid by their greater durability and the heavier cuts which can be taken with them. Some useful information as to the difference between cast steel, blister and shear steel, and their modes of manufacture, will be found in another volume of this series, and at the end of the present chapter a chemical method is given by which the percentage of carbon in any specimen of steel may be determined, but as far as a crucial test of its suitability is concerned, we much doubt the existence of any except actual trial. The fracture of a hardened bar is indeed to some extent a guide, for although uniformity in the grain cannot be accepted as conclusive evidence of its being of good quality, the want of it is a certain indication of its inferiority.

Provided, then, that the tool—whether chipping chisel, drill, or other instrument—has been formed out of thoroughly good cast steel, by careful forging at the lowest possible heat, the hammering having been continued equally throughout the cutting portion till the metal has become almost cold, attention to the following description will probably enable the process of *tempering*—in its reference to the second or ordinary workshop use of the term 'temper'—to be successfully carried out. If, however, these points have been disregarded, good results cannot be expected. The well-known properties which enable this process to be applied to steel, and by which it may be readily distinguished from wrought iron, are these :—First, on being heated and suddenly cooled it is rendered hard and brittle instead of being soft, ductile, and inelastic, as it is when allowed to cool slowly; and secondly, if, when in the hard condition it be again heated, it loses more or less of its hardness according to the temperature to which it is raised, elasticity at the same time taking the place of brittleness. Our object, therefore, in tempering tools is to obtain the greatest possible amount of hardness without sacrificing to too great an extent the requisite toughness and elasticity, for which we must have recourse to the two distinct processes of *hardening* and 'tempering,' or *letting down* the temper. The first of these is effected by heating the tool, or the portion of it which is to be tempered, to a point not exceeding 'cherry red' (for beyond this the heating should never be carried either in forging or in hardening), and on its withdrawal from the fire instantly plunging it into a vessel of cold water. If the heat has been sufficient, the tool will now be found to be exceedingly hard, sufficiently so to scratch glass readily; if not, it must be re-heated to a slightly greater extent, but the lowest available temperature should always be employed. The next step will be to temper or let it down, and for the sake of simplicity we will for the present suppose that the entire tool is to be treated, although this is not very often the case in the workshop. As above stated, this operation consists in again heating it; but inasmuch as it is of the greatest importance to raise it up to, but not beyond the temperature at which the particular steel under treatment receives the appropriate temper, some indication is necessary to inform us when this has been reached. That on which reliance is generally placed depends upon the circumstance, that the superficial oxidation which takes place when a piece of brightened steel is heated in the air, is accompanied by constantly varying coloration of the surface as the temperature in rising passes from about 430° to 600° Fahrenheit. The colours which it successively acquires, and which are given in the following table—which was arranged many years ago by Mr. Stodart—thus form a valuable index of the temperature, and serve to show the point at which the heating must be arrested.

	Fahr.		Fahr.
1. Very pale straw yellow .	430°	7. Light purple	530°
2. A shade of darker yellow	450°	8. Dark purple	550°
3. Darker straw yellow . .	470°	9. Dark blue	570°
4. Still darker straw yellow	490°	10. Paler blue	590°
5. A brown yellow . . .	500°	11. Still paler blue . . .	610°
6. A yellow tinged slightly with purple	520°	12. Still paler blue with tinge of green . . .	630°

Of these 12 tints, Nos. 1 and 2 approximately indicate the temperature to which instruments for which the greatest hardness is required—such as cutting tools for metal—should be raised; Nos. 3 and 4 those adapted for wood-tools, and for screw-taps; Nos. 5, 6, and 7, those which may be used in tempering saws, and also hatchets, chipping chisels, and other tools which are subjected to percussion; Nos. 8 and 9 for springs; whilst by those which correspond to Nos. 10, 11, and 12 steel is let down to so great an extent as to render it altogether too soft for the above purposes. To enable these colours to be observed, the surface of the steel must be brightened upon the grindstone after hardening, and in such a case as we have been supposing, where a considerable portion has to be brought to one uniform temper, great care will be necessary to perform both this and the previous operations of heating and cooling with sufficient uniformity. In this the following general directions may be of assistance, but the particular treatment must, of course, often be varied according to the form and size of the object to be tempered.

In the first place the 'scale,' which forms upon the exterior of a steel forging as well as upon an iron one, should be removed by grinding or filing up the entire surface of any portion over which the temper is to extend, in order that it may receive and part with its heat equally throughout. Secondly, in being hardened, the object should have ample time to become equally hot all over, by being allowed to 'soak' in a fire of sufficient size to surround it completely. For all purposes connected with steel, a charcoal or coke fire is preferable to one made of raw coal; but in hardening, contact with the fuel may be prevented by inserting an iron tube or box into the fire and placing the instrument to be heated within it. When it has thus been brought to the lowest red heat at which it is found possible to harden the particular sample of steel which is being treated, it should be at once plunged into water—vertically, if its shape admits of it—and remain in the water until it is quite cold throughout, since it is then rather less liable to be cracked or distorted by unequal contraction—accidents which very frequently occur at this stage, and against which it is almost impossible to guard with certainty. Lastly, in the final process of tempering, the same precautions must be taken to ensure uniformity of heating; but as a much lower temperature is required, and its progress has to be carefully watched, the object may be conveniently laid on the top of a clear fire, or if small, upon a heated iron bar or plate.

The first of the tints mentioned above can only be perceived by comparing the object which is being tempered with the brightened surface of a piece of cold steel; and when very great hardness is required, it may be necessary to use a still lower temperature for letting it down. In such cases, since the colour of the surface does not then form a guide, recourse must be had to other methods; immersion in a bath of oil, mercury, or some fusible alloy being amongst the best of them. In tempering cutlery, &c., on a manufacturing scale, this mode of heating is largely used both in hardening and in letting down, the heat of the bath being known either from the melting point in the case of the metals and alloys (except mercury), from the amount of smoke driven off in the case of oil, &c., or by introducing a thermometer into the bath.

The smoking and the flashing temperatures of oil are also frequently turned to account in tempering, without employing a bath at all. This method is known as *blazing off*, and consists in coating the hardened object with cold oil, either by dipping or otherwise, and then warming it uniformly over a clear fire till a copious smoke is given off. According to the degree of hardness required it is then either removed from the fire or the heating is continued till the oil inflames. The temper thus obtained is low, but this method—although not so well adapted for the tempering of cutting tools, may be often used advantageously for steel springs, &c.

But in most of the tools to which the mechanician is likely to be called upon to apply the above processes for himself, it is unnecessary, and even undesirable, to extend the temper beyond the immediate vicinity of the cutting edge. This enables both the hardening and the tempering to be performed with one single heating, in the following manner:—The chisel or other tool is brought to a cherry-red heat, either throughout, or at its cutting end only, according to its length. On being taken out of the fire, this end is instantly dipped, for a short distance only, into the water-trough, being withdrawn as soon as this part of it has been thoroughly cooled. The stem, which has not been immersed, of course retains a considerable portion of its heat, and this, owing to the conducting power of the metal, soon begins to return to the part which has been cooled. On brightening a small portion of the surface (by rubbing it upon a piece of sandstone), the faint yellow tints soon begin to appear, gradually extending themselves downwards to the extreme edge of the tool, and deepening in colour. As soon as the edge is of the proper tint, the operation is brought to an end by immersing the entire tool, so as to deprive it of its remaining heat; it is then ready to be sharpened upon the grindstone. In hardening, care should be taken to keep the tool in motion during its partial immersion, since if this be not done it is liable to be much weakened—sometimes even to break—at the water-line. It is also advisable to dip it vertically in the first instance, especially if any considerable length of it has to be hardened. But for practical information on the subject of tempering, the reader cannot do better than to refer to Mr. Ede's useful work on the treatment of steel.'[1]

The following is Professor Eggertz's coloration-test for determining the percentage of combined carbon in pig-iron

[1] *The Management of Steel*: By George Ede. 4th ed.: London, 1866.

or steel, as given by Dr. Percy;[1] in whose valuable work methods for the determination of phosphorus and sulphur will also be found.

Sesquioxide of iron dissolved in nitric acid, if not too concentrated, yields a solution which is free from colour, or has only a feeble greenish tint. When pig-iron or steel is acted on by nitric acid, the solution is coloured by the carbon product in proportion to the amount of combined carbon present, and there is no action on the graphite. A normal solution is prepared by dissolving some cast steel containing a known amount of carbon, with certain precautions to be described, in so much nitric acid of 1·2 specific gravity, that every cubic centimetre (=0·037 fluid oz.) of the solution may represent 0·001 grm. (=0·015 gr.) of carbon. This normal solution does not maintain its colour, but generally becomes paler after twenty-four hours. Feebly burnt sugar gives a yellow, and hard burnt sugar a brown solution; by dissolving a mixture of the two in a solution of equal parts of water and alcohol, it is possible to obtain a yellow-brown normal solution of the proper tint, which may be kept for some time in an hermetically-sealed tube pretty well protected from the influence of light. In order occasionally to control the normal solution, 0·1 grm. (=1·543 grs.) of steel containing a known weight of carbon is dissolved in 5 cubic centimetres (=0·18 fluid oz.) of nitric acid, and the solution diluted until the tint corresponds to that of the normal solution of burnt sugar.

The process is conducted as follows:—0·1 grm. (=1·543 grs.) of the finely divided iron or steel is put into nitric acid of 1·2 specific gravity, and free from chlorine, contained in a test-tube of about 4 inches in length and 0·4 inch diameter. The test-tube is immersed in water, and kept at a temperature of 80° C. If the temperature exceeds this, the

[1] *Metallurgy, Iron and Steel*: By John Percy, M.D., F.R.S.: London, 1864.

colour of the solution decreases and shows too small an amount of carbon. By a lower temperature, the dissolving proceeds too slowly, and the colour of the liquid may be too strong.

When the evolution of carbonic acid gas ceases, which for steel usually requires two or three hours, the test-tube is removed and left to cool. The solution is then carefully decanted off from any black particles which may have been deposited during cooling, into a graduated tube. A few drops of nitric acid are added, and heat applied. If no evolution of gas occurs, the black particles consist of graphite or slag; if otherwise, the test-tube is cooled and the solution is added to that before obtained, and the whole diluted with water until the colour corresponds to that of the normal solution.

If 1 cubic centimetre of the normal solution correspond to 0·1% of carbon, and the solution in the graduated tube measures 7 cubic centimetres, then the iron or steel operated upon contains 0·7% of carbon.

As it is usually difficult to dissolve 0·1 grm. (=1·543 gr.) of iron in less than 1·5 cubic centimetre of nitric acid, it is not possible with the before-mentioned normal solution to determine less than 0·15% of carbon. When the carbon exceeds 0·5%, the solution has a greenish tint, which causes some little difficulty in comparing it with the normal solution; in such a case, a poorer normal solution is made by adding 6 parts by measure of water to 3 parts of the common normal solution. If the quantity of carbon is large, as in white pig-iron, only 0·05 grm. (=0·74 gr.) is to be employed.

In several ironworks in Sweden where the Bessemer process is practised, this method for determining the carbon has already been adopted, and has afforded much facility and certainty in the assortment of the steel, which before it was necessary to test experimentally by forging and hardening.

We may add that the amount of combined carbon in steel ranges from about 0·25 to 1·00 per cent.; that which contains a large proportion being much more suitable for edge-tools than that in which the percentage is small. The former, however, requires both more careful heating and more labour in working. But although a knowledge of the proportion of combined carbon is of very great assistance in enabling us to form an opinion as to the quality of steel, it is by no means an infallible guide. The presence of very minute quantities of other substances—notably of sulphur and phosphorus, and in all probability of silicon also—is capable of exerting upon it great influence for evil; and, on the other hand, it seems to be quite possible that other elementary bodies may have equal effects which are beneficial. At present the general aim of manufacturers is to produce a combination between the purest iron obtainable and a definite percentage of carbon. For doing this various 'direct' processes have of late years been devised (by Bessemer, Siemens, and others), which, although they are admirably adapted for producing steel suitable for many purposes, cannot compete, as far as cutting-tools are concerned, with those made by the old processes of cementation and casting. The mechanical treatment which the finished steel undergoes is also in the highest degree important; and a very recent improvement of Sir Joseph Whitworth in this direction—that of compressing cast steel while in its fluid state—bids fair completely to revolutionise the final stages of its manufacture.

CHAPTER IV.

ON THE FORMATION OF STRAIGHT-EDGES AND SURFACE-PLATES.

BESIDES the measuring instruments and the cutting tools already mentioned, various forms of apparatus are required to enable the worker in wood or in metal to test the accuracy of his operations. For this purpose the former has recourse to the straight-edge, square, bevel, &c.; while the latter, in addition to these, uses templates of various kinds, and—above all—the planometer or *surface-plate.* The importance of this last instrument is so great that we shall devote much of this chapter to the consideration of it; especially as the successive steps in its production will afford a good illustration of the modes of using several of the wood and metal tools previously noticed.

In the year 1840 Sir J. Whitworth called the attention of the scientific world to the complete inefficiency of the process of grinding, by which alone it had up to that time been attempted to produce truly plane surfaces in metal; at the same time exhibiting some plates, the accuracy of whose surfaces very far exceeded the limits previously attainable. They were in fact the first which were worthy of being called true planes, and the circumstance of having discovered and successfully carried out the means of producing them, would alone be amply sufficient to immortalise a name which has already so frequently appeared in these pages.

It is proposed to give in some detail the complete course which must be followed if it be required to 'originate' a surface-plate—that is to say, to produce one without the assistance of an existing surface-plate. Its successful accomplishment is perhaps the highest triumph of mechanical manipulation, to which the reader must not expect to find any royal road. We can do no more than give him a few directions in the difficult path by which he may arrive at success, if he be possessed of a moderate amount of skill, coupled with indomitable perseverance. To give some idea of the agreement which exists between two perfect surfaces, it may be mentioned that if the perfectly clean and dry faces of a pair of Sir J. Whitworth's surface-plates be placed one upon the other, the upper one will appear to float upon a thin stratum of air which for some time supports it without its being anywhere in contact with the lower plate. Conversely, if this stratum of air be expelled by pressing and sliding the plates together—on lifting the upper one the lower one will also be lifted, and may thus be supported for some seconds. The cause of this in each case is that the weight of the plate—although by no means inconsiderable—is insufficient to overcome the resistance which the air encounters in passing through the minute and uniform space which separates the surfaces.

Before commencing upon the cast iron or other plates whose surfaces it is intended to convert into true planes, two preliminary steps must be taken; in the first place two wooden straight-edges must be prepared; secondly, with the assistance of one of these, a set of three steel straight-edges must be worked up. When these have been made to agree perfectly with one another the surface-plates may be commenced; the principle of obtaining exact agreement between a set of three of these being always adhered to. In general, of course, the possession of an existing surface-plate, or at least of a fairly reliable straight-edge will obviate the necessity for performing some of these operations; but we prefer to give them in their entirety, noting afterwards those which the ordinary appliances of the workshop would enable us to dispense with.

First then as to the *wooden straight-edges*; the preparation of which will require some skill in the use of the joiner's planes described in Chapter II. Two pieces of hard straight-grained mahogany not less than three feet long and a quarter

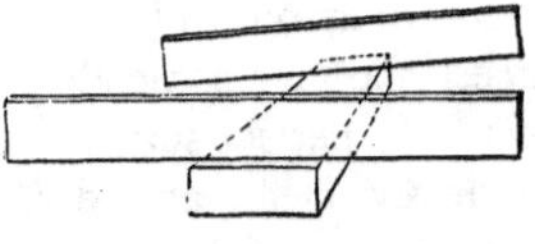

FIG. 83.

FIG. 84.—Marking-Gauge (⅛).

of an inch thick having been selected, and both sides of each having been roughly planed over with a jack-plane, the first care must be to see that they are not 'in winding' —as a joiner expresses it when two of the opposite corners of a piece of wood stand higher than the other two. This is most easily detected by placing two short pieces of wood with parallel edges, called *winding-sticks,* in the position shown in Fig. 83. On bringing the eye nearly into the line of their upper edges, so as almost to make the nearer conceal the more distant one, this error, if it exists, at once becomes visible. After each side has in turn been corrected, one of them should be worked up with a finely-set trying-plane until a continuous shaving can be cut from the full length of the piece. If the plane be in good order, and the caution previously given as to planing 'hollow' rather than 'round' be not neglected, this side will be found to be sufficiently flat.

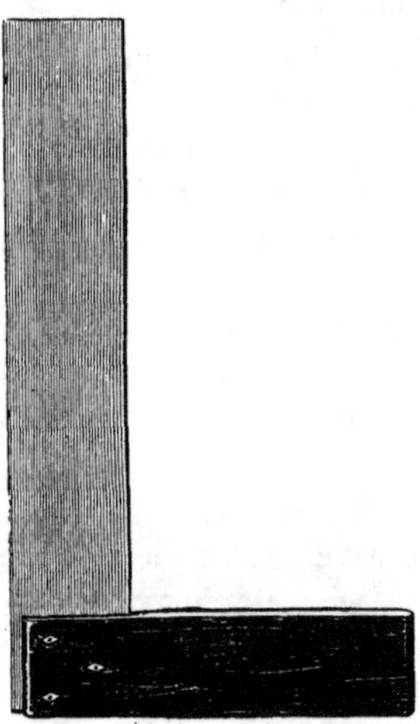

FIG. 85.—Carpenter's Square (⅛).

From this flat surface the desired thickness must be set

off at each end with a *marking gauge* (Fig. 84); the scratch which it leaves indicating the amount to be planed off the opposite side. This done, the edges may be commenced; the process being similar to that by which the sides were made true, but requiring to be conducted with increased care.

The *square* (Fig. 85) will now be necessary to ensure the edges being kept perfectly perpendicular to the surface first obtained—which surface should be marked, and always be worked to. One edge of each piece having been then planed as true as possible, their agreement with one another must be tested; first by observing whether their contact is continuous when they are placed edge to edge and held up towards the light, secondly by clamping them side by side with their true edges upwards and trying with the fingers whether one edge stands to any sensible extent above the other at either end or at any intermediate point. (A still better system is to make three straight-edges, each of which must agree with any other, as described below; but the material not being susceptible of any very great degree of accuracy, it is hardly necessary.) The best result only being obtained when the final cut has been taken from the whole length of the straight-edge, no after-treatment of any kind should be required. But in default of his having the requisite skill, the beginner can with advantage make use of the cabinet-maker's scraper, or—which is much better—a flat piece of cork covered with fine glass-paper.

Fig. 86.—Sliding Bevel (⅙).

Applying the square to the true edge the exact length is now marked off, and the ends cut and planed or 'shot' from the finished towards the unfinished edge. Finally this second edge is marked out with the gauge, planed up, and chamfered if required; the uniformity of the chamfer being checked by applying to it a *sliding bevel* (Fig. 86) which has been previously set to the proper angle.

In this last operation, and also in planing the ends or edges of thin pieces of wood generally, much assistance may be derived from the use of *shooting-boards.* Fig. 87 shows sections of such boards, one of them being intended for planing pieces with square, the other with chamfered or bevelled edges. The trying plane is laid on its side and worked backwards and forwards with one hand, the work being brought up to it with the other.

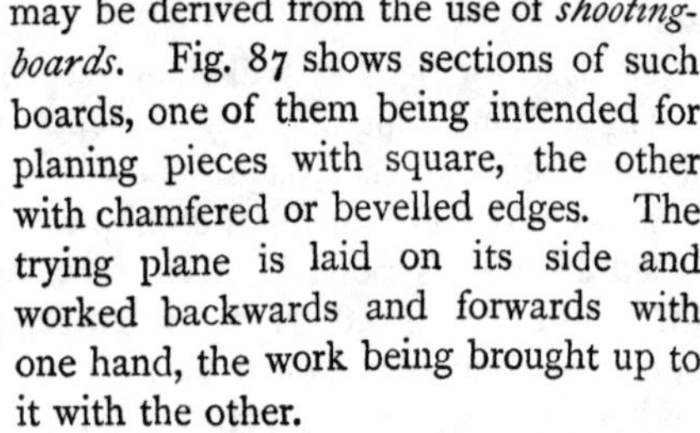

FIG. 87.—Shooting-Boards.

The second step—viz., the preparation of three very accurate *steel straight-edges,* can now be proceeded with. And although we shall confine ourselves to steel, as being the best material for our present purpose (and for straight-edges generally, provided that their length does not exceed about four feet), a somewhat similar course will apply to the working up of longer ones, which would be made of cast iron. The thickness of those which are of steel is usually from $\frac{1}{16}$ to $\frac{1}{2}$ of an inch.

Three similar strips of steel of the desired size having been procured, and their sides filed or smoothed upon a grindstone, holes should be drilled in similar positions near the two ends of each of them. When the strips are laid one upon another, two tightly fitting pins can then be passed through these holes; in which condition the three will resemble a single thick bar. They can be fixed in a vice, and—as far as all the earlier stages are concerned—may be operated upon simultaneously.

The ordinary vice has appeared in a former engraving (Fig. 66). The teeth with which its jaws are filled can in cases like the present be prevented from marking the work held between them, by interposing *clamps*, formed out of pieces of sheet lead or zinc. In filing up the steel straight-edges great assistance will be derived from one of the wooden ones previously made, which should be rubbed with a piece of red chalk, and frequently applied to them; any prominent parts being thus rendered visible. The large rough files used in the earlier stages must be successively replaced by smaller and smoother ones as the work progresses. When it becomes no longer possible to detect any inequalities on applying the wooden straight-edge, the pins connecting the three pieces must be removed, and all further operations conducted upon each of them separately. But before this point has been reached, one edge of each should be bevelled, if required, as the straightness is liable to be impaired by the process.

A systematic course of comparison of one with another must now be commenced, and be persevered in till each edge agrees perfectly in any position with the other two. The necessity for operating upon three (the method above given for wooden straight-edges being insufficiently accurate for steel ones), is evident if we consider that two may agree perfectly together, although one may be convex or 'round,' and the other concave or 'hollow.' When their ends are reversed they will also agree, if the curvature be uniform. But if there are three (which we will call A, B and C), then if A be hollow, B and C when made to agree with it will both be round, so that they cannot agree together. They will in fact make the error appear to be twice as great as it really is, and in order to correct it, an equal amount must be removed from each, until perfect agreement is obtained between them. Then, on correcting A till it agrees with B, the correspondence or the want of correspondence between A and C will prove their mutual truth or error. By persevering and systematic repetition of these comparisons, good results will at length be obtained. And in addition to placing the two which are under comparison with their edges together, and observing whether any light can pass between them, it will be found advantageous to lay them side by side on a flat table or bench and to rub their cleaned edges somewhat forcibly together. Any prominent parts will then receive a slight burnishing, which will show where the file must be applied.

When one edge of each piece has thus been rendered straight, the opposite one may be similarly treated; their approximate parallelism being obtained, when necessary, by the frequent use of callipers during the operation. Of these instruments one example has already been given, and others will be found in a subsequent chapter (see Fig. 110). A simple and very efficient substitute for them may be made by filing a parallel-sided notch, of the exact width required, in a piece of sheet metal. This forms a temporary gauge of no mean accuracy when carefully used; a second notch of slightly greater width, to show when the desired limit has nearly been reached, being a useful addition to it.

Having now at least one thoroughly reliable steel straight-edge, we are in a position to commence the *surface-plates* themselves. Three hard cast-iron plates are required, and whatever their size may be—for this will of course vary according to the purpose for which they are intended—they must have no tendency to bend or become distorted during their preparation, or in after use. To fulfil these conditions it is necessary to support their under sides by deep main and cross ribs. If, however, the main ribs were to follow the rectangular outline of the plate, there should be a support at each of the four corners, and any inequality of the bench or other surface on which it was placed would make these bear unequally, and would be liable to distort the plate. Sir J. Whitworth therefore devised the plan, now universally adopted, of reducing the number of supporting points to three, and arranging the main ribs between them in the somewhat triangular form shown in Fig. 88. Upon the three supporting points the roughed-out castings should rest for two or three weeks before the finishing processes are commenced, in order to allow the plates to settle in the form which they will naturally assume.

Supposing a planing machine not to be accessible, the upper surface of each casting must be prepared with a chipping chisel and hand file, during which operations the chalked wooden straight-edge will render much assistance in pointing out the more prominent portions, just as it did in the preparation of the steel straight-edges. It must of course be applied to all parts of the surfaces, diagonally as well as parallel to the edges. Or a rectangular piece of hard wood planed as true as possible, may be substituted for it with some advantage. But either of them must soon be rejected in favour of the much more accurate steel straight-edge, which can either be held so as to show any depressions by the passage of a streak of light between its edge and the surface, or be made to redden the projecting parts by smearing it with a mixture of red ochre and oil. Moreover, inasmuch as files of considerable length only can be used, which cannot be made to remove small portions from particular spots, as can be done in the case of a narrow straight-edge, to which very light and smooth files can be applied, it soon becomes necessary to substitute a scraper. With this tool very small quantities of metal can be taken off at any required point, so that a tolerably flat surface will soon be obtained. Fig. 89 shows the method of using this invaluable instrument.

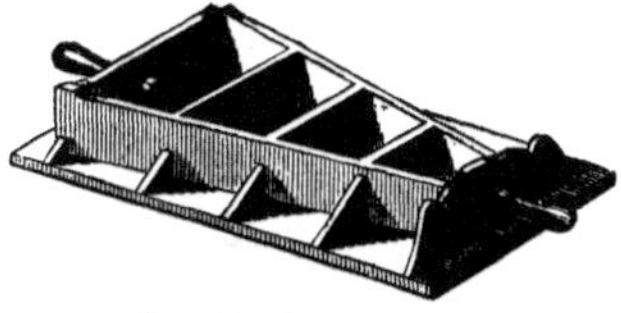

Fig. 88.—Surface-Plate.

One of the castings only should be worked up to the highest state of perfection obtainable with the steel straight-edge and scraper, the care bestowed upon it being repaid with interest during the subsequent processes. This plate we will call A, the others respectively B and C. The face of A is now thinly coated with ochre and oil, B and C being successively made counterparts of it by repeatedly placing them face downwards on A, and then lowering with the scraper all the reddened points on which they bear. This must be continued till the contact between A and B and A and C is as perfect as possible, which will be known by the bearing points—at first perhaps only two or three in number, and of

FIG. 89.—Method of holding Scraper.

small size—having extended themselves so that the entire surface becomes reddened.

A series of comparisons exactly similar to that described in the case of the steel straight-edges, must now be carried out :—B and C being compared together, and corrected by removing an equal quantity from each; A being then made a counterpart of B, and A and C compared with one another. If B and C have been perfectly equally reduced (and to effect this the greatest care will be well bestowed) their errors, which are equal in amount and similar in kind, will have been entirely got rid of, and A will be found to agree with C, thus proving all three of them to be true planes. But it is not likely that this result will be obtained till the above routine has been many times repeated, in doing which the same order should always be adhered to. As the process approaches completion increased watchfulness will be necessary, so as to guard against the introduction of fresh errors; the penalty for scraping off the slightest excess from any one part being the performance of the difficult task of lowering the entire surface to exactly the same extent.

When all possible care has been taken, it must not be expected that the surfaces produced will be absolutely true —mathematical accuracy probably could not be maintained even if it were possible in the first instance—but the approximation to it will be sufficiently close for every practical requirement. The abandonment of the old process of alternately grinding together with emery-powder and water the faces of a set of surface-plates, has been accompanied not only by the most marked improvement in their accuracy, but also by a diminution in the time expended upon their production.

Soon after he had thus successfully produced his surface-plates, Sir J. Whitworth entirely abolished in his own workshops the practice of grinding metallic surfaces together, and although his workmen were considerably prejudiced in favour of it, a very short experience taught them that scraping was by far the quicker and more efficient process. Two considerations point to the wisdom of this step:—first, that any abrasive powder so used cannot but be irregular and uncontrollable in its action, owing to the tendency of the powder to dispose itself unequally over the surface, and also to the fact that in grinding, as generally practised, some parts are exposed to the friction for a much longer time than others; secondly, that any slides or moving parts of machinery so treated, unless when finished they can be entirely freed from the powder (which is almost an impossibility), become the instruments of their own destruction.

Three surface-plates having been successfully 'originated' in this manner, two of them should be kept with scrupulous care for the correction or production of copies of sufficient accuracy for ordinary workshop purposes. Once in possession of a reliable standard, the process of copying it involves a comparatively small tax upon the patience and skill of the operator. Not only is he no longer dependent upon an accurate straight-edge—he may, if he can avail himself of a planing-machine, dispense with the whole of the roughing-out process above detailed. With these invaluable machines—which are described in the sequel—the surface of the rough casting can be brought to such a comparatively finished state, that it may be at once compared with the standard surface-plate. This, being kept thinly and uniformly reddened, will show the prominent parts, to which the application of the smooth file, and subsequently of the scraper, must be confined, as already mentioned.

Long straight-edges, which—as stated above—are generally made of cast iron, require to be so constructed as not to be liable to flexure from their own weight when in

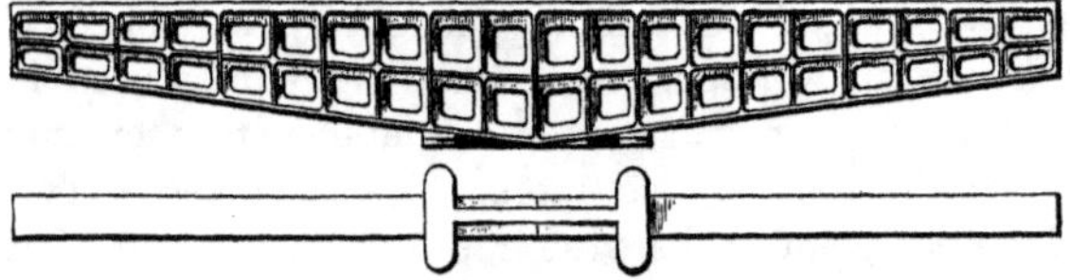

FIG. 90.—Six-feet Cast-iron Straight-Edge.

different positions, nor from that of any articles which may be placed upon them for comparison. The strongly-ribbed form of those made by Sir J. Whitworth will be intelligible from Fig. 90, in which one of six feet in length is represented in elevation and plan. The width of its face is $2\frac{1}{4}$ inches; cast iron straight-edges being always much wider than steel ones in proportion to their length. Those which are 10 feet long have a width of as much as 3 inches, so that they resemble elongated surface-plates, and their preparation requires much greater time and care than is necessary in the case of those of small width with which we have been dealing. For with regard both to straight-edges and surface-plates, it will be readily understood that any given length or area is worked up with far greater facility when it does not form part of a much greater length or surface, than when it does. On this account surface-plates for ordinary work are less advantageously made of a square than of an oblong form—2 : 3 being the most frequent proportion between the width and length. For special purposes, however, surface-plates of special shapes and sizes are often required, and for extending a plane of great area, a good sensitive level is used in conjunction with them.

The use of a good surface-plate to the worker in metal is sufficiently obvious. It enables him, either by following the whole process for obtaining a true counterpart of its surface, to give almost perfect accuracy to the flat facets which occur on the exterior of his work, or by completing the early stages only, to render them perfectly free from twist and prominences, though not necessarily from depressions. A good fit may thus be ensured between the several portions of a machine which are to be attached to one another, with an amount of labour which bears but a very small proportion to that required for 'facing' a slide-valve, or any other moving parts. When several surfaces of moderate size have to be worked up on the same casting or forging, it is advisable in the first place to rough out the whole of them with moderate accuracy, afterwards finishing them in the order of their size. The largest, from which the correct position of the other surfaces should almost always be determined, will thus be first completed. If there be a second surface parallel to it, this should next be marked out, either with a gauge of similar construction to the wooden marking-gauge shown in a previous figure, or with a *surface-gauge* (Fig. 91), used on the surface-plate. In a similar manner the truth of the edges—if they be at right angles to the face—can be tested conveniently by placing the work, face downwards, upon the surface-plate, and applying to it the exterior angle of a steel square, which for this purpose should have a wide back, so as to enable it to stand upon the surface-plate.

But these methods of forming flat surfaces either parallel or at right angles to other existing surfaces are available only when no very high degree of accuracy is required. In cases

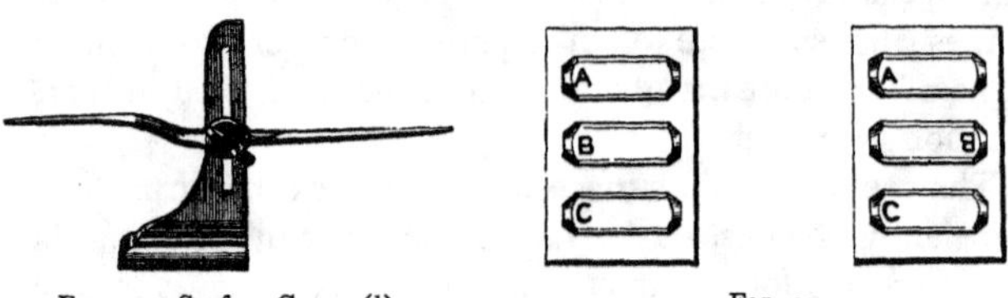

FIG. 91.—Surface-Gauge ($\frac{1}{8}$). FIG. 92.

for which they are insufficient it will be necessary to follow them up with treatment of a very different kind.

We will suppose, for instance, that three bars of rectangular section are to have their sides and ends worked up sufficiently truly plane and parallel to enable them to be used as standard measuring bars; and less than three cannot be successfully treated. With the assistance of one of a pair of surface-plates—of which the size must be sufficient to admit of the three bars being placed upon it simultaneously—one side of each bar is made as truly plane as possible, and upon these finished sides the three are then laid in the position shown in Fig. 92. The opposite sides, which will of course have been made nearly flat and parallel in the preliminary working up to which all parts of the bars have been subjected, are then tested by applying to them the other surface-plate, reddened; their prominent parts which receive the colour being successively lowered in the ordinary way, till the whole of the surfaces are proved, by being uniformly coloured, to be in one and the same plane. If this plane be perfectly parallel to the surface upon which the bars are resting, they will be equally and uniformly coloured on applying the upper surface-plate, after turning the central bar (B) end for end (as shown in the figure), or after changing the places of A and B, or B and C. But if each bar either has one of its edges or one of its ends higher than the opposite one, the reddened plate will no longer take its bearing upon the entire surfaces; and the treatment must be continued—scraping away the coloured portions, then changing their positions from side to side, and reversing them end for end—until the entire surface is reddened, in whatever position the bars may be placed. This, as might be expected, is often a work of very great labour, especially when, after making them parallel, one of these sides has to be uniformly lowered in each bar in order to make the thickness correct. This thickness can only be satisfactorily tested with the aid of a measuring machine.

The next step will be to make one of the remaining sides of each bar perfectly perpendicular to the two sides which have just been rendered parallel, and the method of doing this will be intelligible from Fig. 93. One side (we will say of the bar A) is first brought to a true plane in the ordinary way. A second bar (B) is then similarly treated, its exact correspondence with the first being ensured by frequently comparing them together in the manner shown in first portion of the figure, the side of A being reddened to show the extent to which the two are in contact. When their agreement is perfect, one of the bars must be reversed; the effect of which will be, if their former plane of contact was perpendicular to the horizontal surfaces, to leave their agreement as perfect as before, but if not, to make the error appear to be twice as great as it

FIG. 93.

really is, as is represented (greatly exaggerated) in the second half of the diagram (Fig. 93). Equal amounts must then be removed from the prominent portions of these sides until their contact is found to be perfect in either position. One side of the third bar can be made correct by testing it with either of the other two, after which the fourth side of each must be made parallel to that which is opposite to it, and the thickness corrected as before.

We have thus worked up the whole of the sides to the condition of plane surfaces truly at right angles to one another, the bars being now "die-square," as it is termed, throughout their length. It only remains to reduce to the

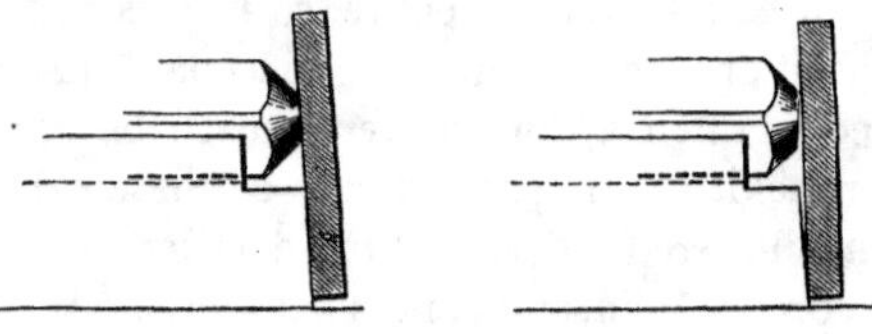

FIG. 94.

same conditions the end surfaces—which in the bars under consideration are turned down from a square to a circular form. For this purpose there will be required, in the first place, a small surface-plate, called a 'trial plane,' and secondly, a cast-iron bed, having on its upper surface a V-groove, of which the sides are truly plane, and at right angles to each other. The ends of the bed must also be made approximately perpendicular to the direction of the groove. One of the bars being laid in this groove—and throughout this part of the process each is treated quite independently of the other two—the trial plane is applied to the end of the bed in the manner shown in Fig. 94. If it is found that whilst the reddened trial plane is held firmly in this position, a portion only of the end surface of the bar is coloured by it, scraping must be had recourse to and be continued till coloration of the entire surface proves it to be in one and the same plane with the extremity of the bed. On now laying the bar upon those which before were its upper sides, perfect coincidence with the end plane of the bed will obtain only if this plane be truly perpendicular to the axis of the bar. Any departure from this direction will be rendered evident by a portion of the surface only being coloured, owing to the imperfect contact which the right hand portion of the above figure so plainly shows. This must be corrected by scraping the ends of both the bar and the bed.

When the sides or edges of a piece of work are inclined instead of being perpendicular to a previously formed surface, the correctness of the inclination can be tested approximately by means of a bevelled square, set either permanently or temporarily to the proper angle, and used either with or without a surface-plate. But a more usual method is to prepare a *template*, by cutting out a piece of stout sheet metal to the converse of the required form. Templates are also much used in working up curved or irregularly shaped surfaces. When many of these have to be worked to the same curve or angle, a pair of templates, called respectively a male and female template, is generally made in the first instance. Each of these is the counterpart of the other, their perfect agreement being tested by holding them up against the light —none of which should pass between their edges. Any deterioration in the form of the one which is applied to the work can then be at once detected by comparing it with the other. But in the treatment of the flat portions of any piece of work, whatever may be their positions with regard to each other, templates alone should not be relied upon. The relative positions of the surfaces having been established by means of these plates, their flatness should always, if possible, be checked by applying them to a surface-plate.

When the work itself is not readily moveable, owing to its weight or other causes, the surface-plate can be lifted by its handles, and applied to it with almost equally good effect.

Besides the perseverance, which, as we have said, is a first essential to anyone who would put in practice the instructions above given, his success will to a great extent be measured by the amount of skill in the use of a file to which he has attained. A few hints on the subject may, therefore, be acceptable, although the art is one which can be thoroughly acquired by practice only.

Files, inasmuch as they differ from cutting tools generally, in being incapable of being sharpened, demand greater economy in their use. On this account, a new file should always, if possible, be first used upon brass or cast iron, for with these metals a blunt file is comparatively useless; afterwards upon wrought-iron or steel, which do not require similar sharpness. On the same grounds it will be found advantageous in all cases to use as rough a file as circumstances will permit.

The destructive effect of the 'skin' of cast iron upon the teeth of a file has already been mentioned. In the rare cases in which it is necessary to attack it with a file, one which is too blunt for other work should be employed, but it seldom happens that the surface cannot be prepared to some extent, either by planing, chipping, pickling, or grinding upon the grindstone. The last of these processes is much to be recommended on economical grounds—not only for castings from which much of the material cannot be spared, but also for taking off the 'scale' produced by the superficial oxidation of wrought-iron when being forged, which is also to some extent injurious to a file. Brass castings, although they are not hardened by the chilling effect of a sand mould like those of cast iron, must also have the sand which adheres to them removed by pickling or by cleaning them with an old file before a sharp one can be used upon them with impunity. Lastly, the operator should remember that a file will not long continue keen if he allows it to come into contact with the jaws of the vice in which his work is so frequently supported.

The chief cause of the greater difficulty in handling which the file presents to a beginner, as compared with the generality of cutting tools, lies in the small amount of guiding power which it possesses in proportion to the accuracy of the work demanded of it. Since its whole surface consists of a series of cutting edges or points, it is evident that those which are in action at any one time will perform equal shares of the work only when a perfectly equal 'feeding' pressure is applied to each of them. In the due proportioning of this pressure lies the whole secret of success.

Let us consider how this must be effected when it is required to file upon a piece of metal of small width a face perfectly flat and parallel to its existing face. Its section and that of the file teeth in contact with it—both much enlarged—are represented in Fig. 95. In order that the successive cuts may be straight and parallel, it is necessary that the pressures A and B be constantly equal, an equal quantity of the material being then, and then only, removed by each tooth. But — assuming the handle of the file to be held in the right hand and its point in the left—during each stroke the leverage at which the pressure applied by each hand is acting, is constantly varying; the right hand having largely the advantage at the commencement, and the left at the end of the stroke. So the effect of the natural tendency to equalise as much as possible the pressures applied by the two hands, is, at first to force into the work the teeth on the side B, almost neglecting those at A, gradually to diminish this inequality on approaching the centre of the stroke, and again to increase it in the

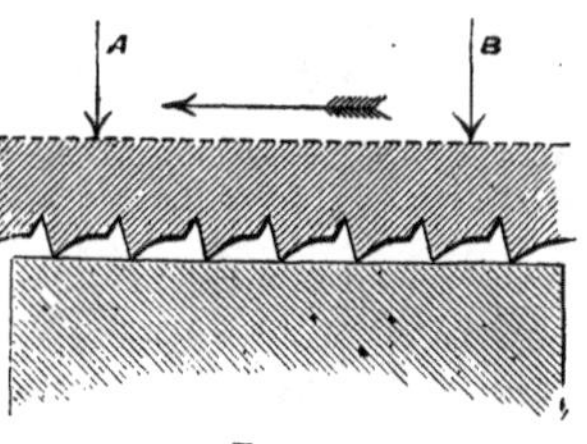

FIG. 95.

opposite direction, after the central point has been passed. Consequently, as is shown in an exaggerated manner in Fig. 96, the side B at the commencement, and the side A at the end of each stroke, will be filed away to too great an extent—a convex surface being thus produced instead of a flat one. Until he has completely overcome this tendency, which practice alone will enable him to do, the beginner should not rest satisfied with his performance. By carefully feeling for the position at which the file bears evenly upon the whole surface which it covers, and checking its disposition to tilt upon either edge at any part of the stroke, he will gradually acquire the habit of moving his hands in the nearly horizontal line required, and so varying the pressure applied with each that it shall be constantly uniform on all parts of the work.

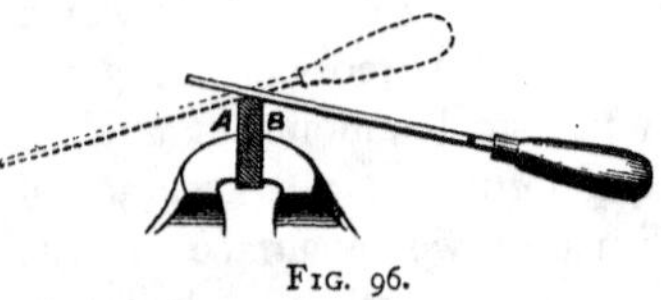

FIG. 96.

The case of filing a narrow facet, such as we have been considering, although it affords a good illustration of the difficulty of handling a file, is not to be recommended for practice till considerable success has been achieved with rather wider surfaces. For this purpose the best width is from about one inch to four inches, according to the size of the file and the capability of the operator. But whatever its size, the work should always, if possible, be so fixed that the surface which is to be filed shall be horizontal.

Wide surfaces—those of surface-plates, for instance—present other and at least equal difficulties. It then becomes necessary, as we have seen, to localise the action of the file, which of course would be impracticable in filing a flat surface with a flat file. Advantage is therefore taken of the convexity which to a greater or less degree is given to almost all files—partly to enable them to be used in this manner on large surfaces, but more especially on account of the great difficulty which exists in maintaining the exact form of the 'blank' during the processes of cutting and tempering. (The necessity for guarding against thus making either side concave, is obvious; inasmuch as it could then never be used upon any flat surface.) To apply the pressure to a file of but slight convexity in such a manner that it shall be in contact with the work only in the part intended, will be found to require no inconsiderable skill, owing to the very small amount of guiding power which it possesses. Short strokes, with the application of local pressure by placing the fingers of the left hand on the back of the file just over the spot to be lowered, are often advantageous; but in order to become a master of the art, the reader must submit to the tutorship of experience, for which verbal instructions are an utterly inadequate substitute. In whatever manner the file is held, it should be almost relieved from pressure during the backward stroke, in order that the teeth may not be blunted or broken by forcible contact with the work whilst they are placed unfavourably for offering resistance to it.

With regard to the height at which it is best to fix a piece of work which is to be filed up, that of the elbow of the operator will form an approximate guide; heavy work being somewhat lower with advantage, so that more of the weight of the body may be brought to bear upon it, and small work being more convenient when rather higher. In the latter case the height is generally regulated by that of the tail- or bench-vice, into which a block of wood for supporting any work of irregular form can be so easily fixed. Thus, when it is required to reduce with the file the diameter of a wire or other small cylinder; a piece of wood being fixed in the vice, a notch is filed in it, and the wire, supported in this, is turned with the left hand through about a third of a circle during each stroke. When the file, as in this case, is held in the right hand only, the requisite pressure is frequently given by extending the forefinger along the back of it—as in holding a carving knife. But into the consideration of the treatment of curved surfaces, either convex or concave, we will not here enter, for they will present no insuperable obstacles to anyone who has become sufficiently adept to produce a true flat surface.

There is no fixed position in which to stand while filing up a piece of work; that can only be determined by the character of the material and the surface to be filed. Thin, and also very small, surfaces are liable to rip the teeth of the file; when convenient to the manipulator, a thin edge may be prevented from injuring the teeth in some measure, by taking up such a position that the file may lie directly along the work rather than directly across it.

In bringing any piece of work to a finished surface by using files of successive degrees of fineness, care should be taken entirely to remove the teeth-marks of the preceding file before passing on to a smoother one. The practice of drawing the file sideways along the work, which to some extent prevents its leaving the marks of its teeth, is not to be recommended; nor is that of giving it a rubbing motion, by which the surface becomes covered with curled scratches, instead of the parallel lines which result from a series of straight cuts given with the entire length of the file. When great smoothness is required a little oil may be used with a smooth file, which will diminish its tendency to become *pinny,* from the fibres (in the case of wrought iron, &c.) being torn up and clogging its teeth. As good work cannot be produced with a file in this condition, a wire *file-brush* should be constantly at hand for clearing it when necessary, any very persistent particles being removed with the scriber, or some other metal point.

CHAPTER V.

ON FOOT-LATHES.

In most of the processes which we have hitherto been considering, the work is kept stationary, and the necessary motion is imparted to the tool. In the various forms of lathe—to which we must now pass on—this operation is almost always reversed, the work being made to rotate, whilst the tool is moved to a comparatively trifling extent. The great facility with which moderate truth of figure can in this way be given to cylindrical work, and the endless variety of forms which result from combining the motion of the work with that of the tool, have made the lathe one of the most widely applicable, as it was one of the earliest of all machine-tools. Not only can the hand-turner, in his comparatively simple form of the instrument, produce with some half-dozen chisels, gouges, &c., any number of intricate mouldings, for each of which the joiner would require a separate moulding-plane, but, by giving a perfectly uniform motion to his tool, he is able to cut spirals or screw-threads of any required depth or delicacy. For this latter purpose alone the lathe is quite indispensable to the Millwright or the Engineer, since by means of it—when at least it is provided with the automatic arrangement, which will hereafter be described—he is enabled to produce screws of any required diameter, or of very great length, with an amount of speed and accuracy quite unattainable by the methods previously mentioned.

In its original form the lathe was doubtless a very simple affair, probably differing but little from the old *pole-lathe*, which may still now and then be met with in a village carpenter's shop. Fig. 97 shows the principle of its construction, and renders intelligible how, without any metal-work whatever with the exception of the two dead centre screws and a tool-rest (the latter omitted in the Fig.), it is possible to construct a lathe which would at least render much assistance in the manufacture of a bedpost. It is

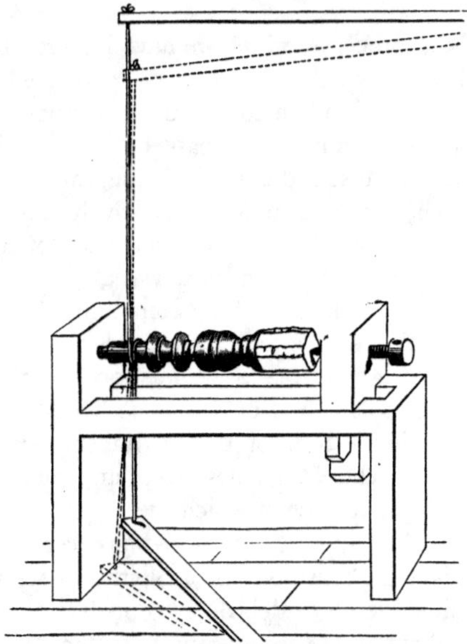

FIG. 97.—Pole-Lathe.

obvious that this machine is not adapted either for accurate or speedy work, the backward and forward motion obliging the turner to waste one-half of his time whilst the spring lath (whence the name *lathe*) is raising the treadle, and thereby producing a revolution in the wrong direction.

The introduction of a crank and flywheel, by which the reciprocating motion of the foot is converted into one of continuous rotation, and the replacement of one of the dead centres by a headstock capable of supporting a mandril with a pulley upon it, were most important improvements, which are believed not to be of very ancient date. But, with the exception of the crank and the mandril, the change did not necessitate the use of much ironwork, inasmuch as the bed, flywheel, pulley, and headstocks could still be made

FIG. 98.—Foot-Lathe.

almost entirely of wood. Lathes of this kind may even now be found in use among soft-wood turners, who, for small work, require a light flywheel, which can be quickly brought up to a high speed, and as quickly stopped. For many years, however, timber has been giving place to metal in the construction of foot-lathes, and it is now rarely employed even for the *beds* or *framing* of those which are intended for accurate turning. For the numerous powerful instruments which the Engineer includes under this head, it is of course wholly unadapted.

Fig. 98 represents an ordinary mechanic's *Foot-Lathe*, the bed (a), the standards (b), the headstock (l), flywheel (i), &c., being of cast-iron; the crank-shaft (h), the chain (f), the framing of the treadle (e), &c., of wrought-iron; the mandril of steel, and its bearings of gun-metal. The only parts of it which are of wood are the backboard (m), and the front of the treadle (e). The conical form of the fly-wheel and pulley enable their relative speeds to be varied; so that this lathe can be used either for wood or brass, either of which require considerable speed and but little power; or for iron or steel, for which these conditions must be reversed.

In the case of soft materials, the size of the work is limited as to length by that of the available part of the lathe-bed, and as to diameter by the height of the centre of the mandril above it. As the diameter of the work cannot be more than twice this height, it is necessary to know it, and also the length of the bed, in order to judge of the capabilities of a lathe. That in the figure would be described as a '5-inch lathe, 3 feet 6 inches bed.'

For the use of amateurs, lathes are made in considerable variety. It is generally required of them that they shall be light in appearance, well adapted for either soft or hard wood, and also for light work in metal. For the first of these purposes it is essential that a lathe should run freely at a high speed, a condition not often fulfilled by those of the class shown in the preceding figure. Considerable gain in this respect results from the abandonment of the parallel bearings of the mandril there depicted, in favour of that shown in the section of the *Fast Headstock*, which occupies the left-hand portion of Fig. 99. At its hinder end this mandril has a steel centre-point, which is supported in the slightly hollowed extremity of a screw, also tipped with steel. A fine leading-hole drilled truly in the centre runs for some little distance into the body of the screw, the tendency of the point to follow this when any wear takes place being of great assistance in maintaining the truth of the centreing of the mandril. The other—the front—bearing, is parallel only for about five-sixths of its length, the remainder consisting of a conical shoulder, which—by means of the screw at the opposite end—is just kept lightly in contact with a steel 'bush' in the face of the headstock. The whole of the bearing portion is case-hardened, and should be so carefully fitted to the bush that, although capable of revolving with perfect freedom, not the slightest tendency to 'shake' in a

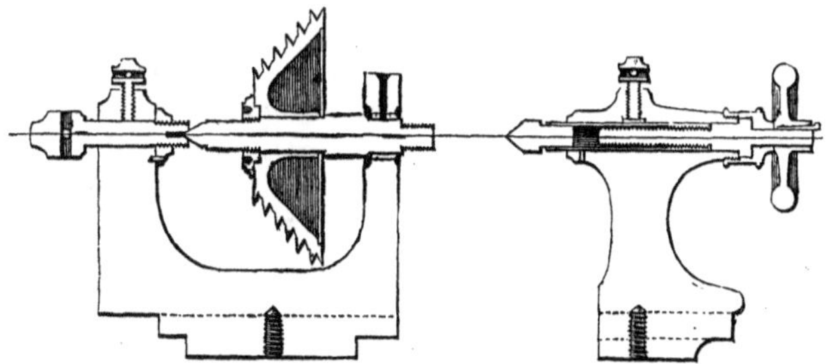

FIG. 99.—Section of Headstocks (⅛).

direction transverse to its axis can be detected. In good lathes of this kind, the conical pulley of the mandril is generally made of gun-metal; and if intended for ornamental *Engine-turning*, its face is graduated with scales of equal parts, into the use of which we cannot enter at present. The flywheel in these is often provided—in addition to the series of grooves round the periphery, which appear in Fig. 98—with an inner ring, having upon it one or more grooves of very much smaller diameter. By passing the lathe-band round this ring and into the largest groove of the pulley, a very slow speed can be given to the mandril.

Besides the fast headstock, the bed of a foot-lathe has to carry a *Rest*, to enable the tool to be held steadily against the work, and also a *Back Poppet*, for holding a steel point or 'back centre.' In Fig. 98 these are marked respectively d and c, but the construction of the latter will be found to be much more intelligible from the section of a Poppet-head in Fig. 99, which, although adapted for a lathe of a superior class, does not differ from the other in principle. The chief use of the poppet is to afford support to any piece of work which would otherwise be unsteady, by giving it a bearing against the back-centre, which is made of steel, and, being pointed, does not interfere with its revolution. It is highly important that the centre should be always exactly in the line of the axis of the mandril, whilst at the same time the distance between it and the headstock must admit of ready and rapid adjustment. The poppet is therefore made moveable along the bed, but its motion is confined to one straight line by one or other of the methods mentioned below; a screw, working into the tapped hole which appears in the section, enabling it to be securely clamped after being brought up to the extremity of the work. The back-centre can then be adjusted, so as to bear with a greater or less pressure, by turning its hand-wheel, which works the sliding cylinder by which it is carried. An examination of the section will explain the reason of its movement. The spindle upon which the hand-wheel is keyed is screwed for about half its length into the sliding cylinder, this being made hollow in order to receive it. But a large shoulder, which is kept up to the back of the poppet-head by a cap, prevents the spindle from changing its own position; and a pin, running in a groove upon its under side, resists any inclination on the part of the sliding cylinder to revolve with the spindle. When, therefore, the hand-wheel is turned, the cylinder is made to slide, towards the mandril when the upper part of the wheel is moved to the right, and away from it when to the left, the screw upon the spindle being made left-handed with this intention. The back-centre having been thus satisfactorily adjusted, the sliding cylinder can be clamped by the set-screw above it.

In drilling metal, &c., with a drill attached to the mandril, a circular plate, flat in front and having at its back a boss fitting into the end of the sliding cylinder, may often be very advantageously substituted for the steel point.

The Rest (d, Fig. 98), requires much greater freedom of motion than the preceding; but it must equally be capable of being firmly clamped in any required position. This is provided for by making a dovetailed groove in the base to which its socket is attached, a bolt, of which the heads fit loosely into this groove, being passed between the cheeks of the bed to the handle (k). By giving half a turn to this handle, so as to screw it a little farther up the bolt, the rest can be instantly made firm, although, as long as the bolt is left slack, it can be moved freely to any position, either parallel, transverse, or diagonal to the bed. Lastly, the height of the T (which forms the actual rest for the tool) can be varied by sliding it into or out of the socket, in the side of which is a set-screw, by which sufficient pressure can be applied to hold it firm. T-pieces of considerable length for long cylindrical work, and of great width upon the face, for supporting the turning tools used for wrought-iron, &c., are generally amongst the ordinary apparatus supplied with a foot-lathe.

The *Beds* of turning lathes generally consist of two rigid beams, or 'cheeks,' set perfectly parallel to one another at a few inches apart. These are in many instances of precisely similar section, as in the lathe represented in Fig. 98, where the upper surface of each is broad and flat. In such cases the poppet is kept straight, at whatever part of the bed it may be placed, by a wide feather, which is cast on the under side of its base, and which exactly fits the space between the cheeks. This feather cannot be seen in Fig. 98, and it is in section, together with the rest of the headstocks, in Fig. 99; but in the latter its depth is shown by the dotted line, which indicates the level of the surface of the bed. The same end may be gained without the employment of a feather, by making the upper edge of one, or both, of the cheeks angular instead of flat, and forming a V-shaped groove (or two, if both the cheeks be angular) in the base of the poppet-head. Another method, which may often be seen in the small lathes used by watchmakers and others, consists in making a single bar of triangular section constitute the bed, the poppet-head being, as it were, placed astride upon its upper edge. The rest-socket then requires to be upon a small sliding table, as indeed it generally is when there are two cheeks, unless their surfaces are of ample width. This form of bed, which is very effective in preventing lateral motion on the part of the poppet, was devised by Mr. Maudsley, and was used by him for large lathes as well as small ones. For the former, however, it has now been entirely abandoned.

Lathe-bands, by which the power is transmitted from the flywheel to the mandril, are usually made of a single length of catgut. The ends are screwed into a pair of small steel sockets, one of which terminates in a hook and the other in an eye, so that they can be connected or separated at pleasure. When bands of different lengths are used on the same lathe, this method of connecting them is advantageous; but when slight variation of speed only is required, as in the case of wood lathes, the following arrangement will be found to reduce the friction upon the mandril bearings, and by consequence to lighten the labour. The diagram (Fig. 100) will almost explain itself, the peculiarity of the system being the introduction of an overhead frame carrying a freely-revolving pulley. One end of this frame is hinged to the wall, the other being lifted by a cord with a weight attached to it. By increasing or diminishing the weight, any desired degree of tightness can be given to the lathe-band; but inasmuch as it takes a whole turn round the mandril pulley instead of less than half a turn, its tendency to slip is so much diminished that it can be kept comparatively loose. Moreover, the tension remains constantly uniform, in spite of changes in the amount of moisture in the atmosphere, which alter the length of a gut band to a considerable extent. A single groove only on the margin of the fly-wheel is required on this system.

The tools of the professional wood

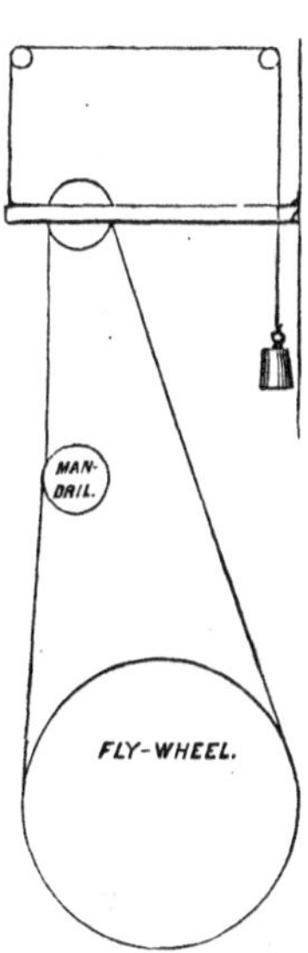

FIG. 100.

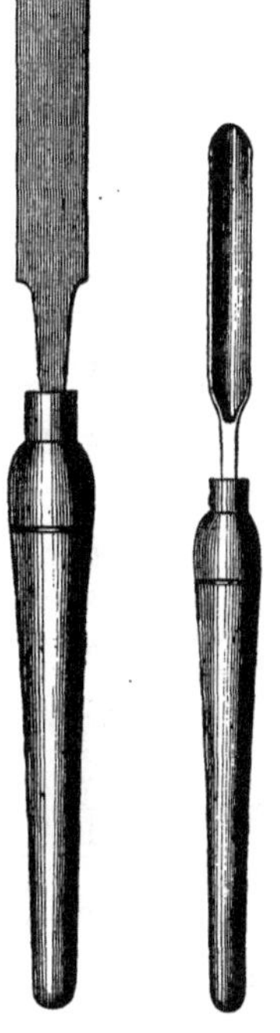

FIG. 101.—Turning Chisel and Gouge (1/8).

turner form a striking contrast, in point of number, to the phalanx which the maker of amateurs' lathes and his customers generally seem to agree in considering necessary. For soft wood, the former is satisfied with two or three *gouges* and *chisels* of different sizes—the larger ones having handles of great length, as shown in Fig. 101—a *parting-tool*, and, for occasional use, a *side-tool*, one form of which is given in Fig. 102. The rounded edge of the gouge, and the oblique doubly-bevelled edge of the chisel, should be noticed, as they differ, both in form and acuteness, from those of the joiners' tools, figured in Chapter II. For hard wood, the turner uses the same gouges, &c., merely grinding them at a more obtuse angle; and he also has frequent recourse to scraping tools, of the same character as the side-tool as regards their edges, but of various sizes and shapes. One of the most useful of these is the *round* tool (Fig. 103), which

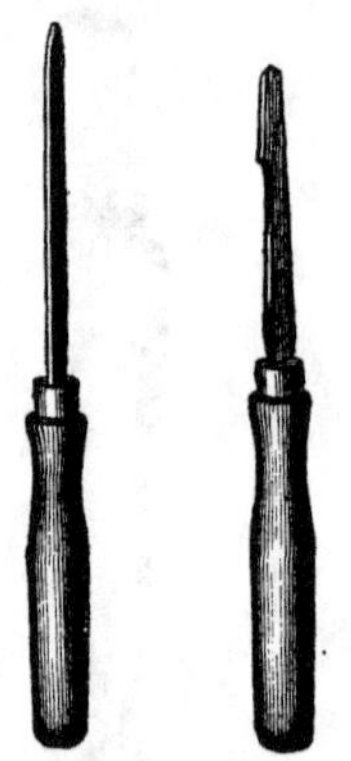

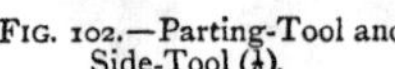

FIG. 102.—Parting-Tool and Side-Tool ($\frac{1}{6}$).

FIG. 103.—Round Tool ($\frac{1}{6}$).

almost takes the place of a gouge for roughing out brass and other hard but not stubborn materials. Various moulding-tools, of more or less complicated forms, are to be met with at tool-dealers', but rarely at turners', shops. The latter generally prefer to use as simple a tool as possible, in special cases altering its form to suit their purpose. A good grindstone, ready for use at any moment, is therefore a first essential to the turner, as it is indeed to everyone whose work depends upon the efficient state of his cutting tools. With soft materials especially, no amount of skill or sandpaper will remedy the evils which result from working with blunt or badly-sharpened tools; and the first lesson which should be learnt by every would-be turner, is the art of using his grindstone and oilstone with speed and effect. The edges of turning-tools which are intended for soft wood should be ground to an angle of from 20° to 30°; those for hard wood, to from 40° to 80°; whilst those which are to be used for brass may even exceed this latter angle, having sometimes a thickness of as much as 90°. In grinding turning gouges, of which the edges possess the double curvature which gives them the circular or elliptical outline seen in the above figure—and the sides may often be advantageously ground away even more than is there represented—the wrist motion

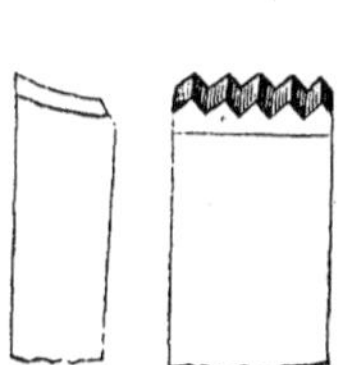

Fig. 104.—Outside Screw-Tool.

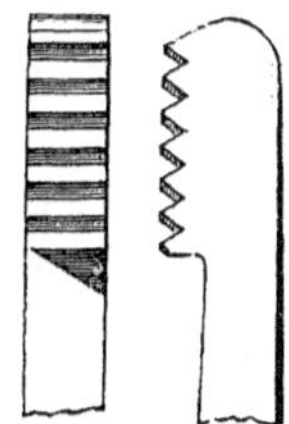

Fig. 105.—Inside Screw-Tool.

mentioned in connection with the grinding of carpenters' gouges is required, and, in addition to it, constant change in the height at which the tool is applied to the stone. In the round-nose and such tools, of which the edges are curved in this one direction only, a similar change of position upon the stone is also necessary; but in their case it must not be accompanied by any revolution about the axis of the tool.

Among the tools for hard wood and brass should be included the hand *chasing-tools* with which screws are cut, chiefly on these materials. Fig. 104 gives the side and back view of an *outside screw-tool*, and Fig. 105 the side and front view of an *inside screw-tool*, the latter being used for cutting female screws. In cutting a screw with either of these, a perfectly uniform motion parallel to the axis of the work must be given to the tool, the speed being such as to make it advance one thread during each revolution of the mandril. As might be expected, the operation requires considerable practice, especially in cutting coarse threads, for which the motion of the tool is considerable—error in the speed imparted to it resulting either in uneven or 'drunken' screws or in double threads. But a good workman can thus produce with ease and certainty screws of the greatest cleanness and delicacy, and—the pressure required being very slight—they can be cut by this method on the thinnest and most fragile materials, which would be quite unable to resist the more violent treatment to which they would be subjected in the screw-cutting processes previously mentioned. This system is used to a very great extent by opticians for the brass fittings of telescopes, microscopes, &c., the thickness of the brass 'triblet' tubes employed frequently exceeding only to a very small extent the depth of the screw-thread which is cut upon them.

FIG. 106.—Diamond-pointed Graver (⅛).

For turning iron—more especially wrought iron—and steel, the preceding tools are not well adapted. For this purpose a slow motion must be given to the lathe, and the edge and inclination of the tool must be such as to give it a cutting rather than a scraping action. This end is tolerably well fulfilled by the *Graver* (Fig. 106), the cut being made with the edge rather than with the point; and it is still more effectually gained by some form of *Heel-tool*, such as that shown in Fig. 107. In using these tools, the 'heel' is supported by the wide and flat-topped T already alluded to; and the pressure produced by the cut is thus transferred directly to the rest-holder and lathe-bed, causing little or no strain upon the hands of the workman. But the necessity for having the tool completely under control, and the small amount of motion to be given to it for determining the depth of its cut, obliges it to be securely fixed in some long and powerful handle. It is therefore used with a *tool-holder*, such as that which appears in the engraving, being laid in a groove on its upper surface, and there held by an eye-bolt which passes through the handle, and which can be tightened by a nut at the lower end of it. The tool-holder is of sufficient length to pass under the shoulder of the operator, its upper portion being firmly grasped by his left hand, and the handle held in his right. The tendency of heel-tools to

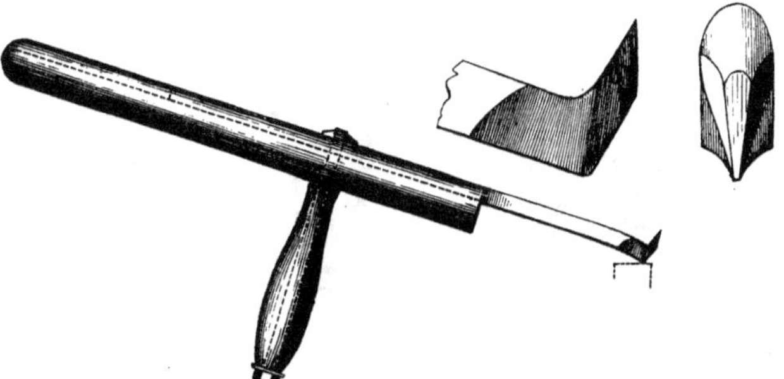

FIG. 107.—Heel-Tool, with details of Cutting end.

catch into the work can thus be resisted; and when once this difficulty has been overcome, they will be found to be easily managed and very effective. Besides these, tools of a few other kinds may be used for hand-turning iron and similar stubborn materials; but since the introduction of the slide-rest their occupation has so nearly gone, that our available space will be more profitably reserved for the consideration of those used with this invaluable instrument.

The very varying nature of the work which a turner is called upon to perform demands similar variety in the methods by which he connects it with the mandril, so that it may share its revolution. For this purpose he provides himself with an assortment of *chucks*, which, being fitted to the mandril by means of the coarse-threaded screw at its extremity, can be readily attached or removed from it. Commencing with those which generally accompany a wood-lathe, we shall find that the one which shows the smallest advance upon the old system of fixing between two dead centres (as in Fig. 97), is the *fork-chuck* (Fig. 108). The fork being driven a small distance into one end of a piece of wood, the other end is supported by the back-centre. For work which requires to be turned on the outside only, without being 'hollowed' at all, this method of fixing it has much to recommend it, especially when the length greatly exceeds the diameter. For rather harder

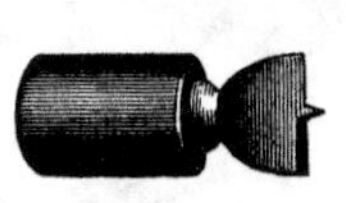

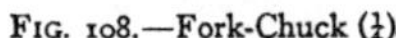

FIG. 108.—Fork-Chuck (¼).

FIG. 109.—Cup-Chucks (¼).

materials a smaller kind of fork is often used—the Cross Kerf—of which the prongs are arranged thus (+), enabling work of much smaller diameter to be turned up to the end without fear of their coming into contact with the edge of the tool. In either case the fork is fitted, either moveably or permanently, into its chuck, which may be either of wrought or cast iron, brass, or gun-metal. The latter material is generally used for this and also for the *Cup-chucks* (Fig. 109), which are supplied with the more expensive wood lathes; but it seems to possess no advantage, at least over malleable cast iron.

Of all the chucks employed by the wood turner, cup-chucks are by far the most generally useful. From three to six of them, of different sizes, are usually supplied; and a few shillings spent in increasing the number will often prove to have been well invested. With the details of the process of 'chucking' we have not here to deal; but the knack of speedily obtaining a thoroughly good fit between the wood and the chuck is one which it is most important for the turner to acquire. In doing this a handy workman will only in rare cases have recourse to his callipers, although in obtaining a fit between two portions of his work, or in working from a pattern of which a correct copy is required, they will be in constant requisition. Two pairs of turners' callipers are shown in Fig. 110, one of them having points at each end of the legs, of which the distance apart is constantly equal, so that the agreement between an outside dimension measured with the one and an inside dimension measured with the other can be easily verified.

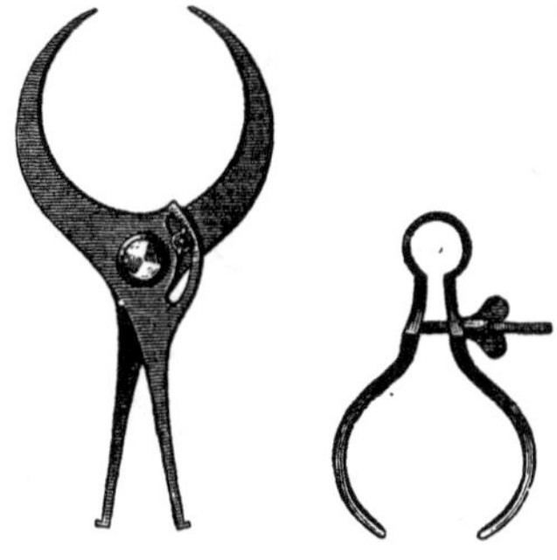

FIG. 110.—Callipers (½).

Once well fixed in a cup-chuck, the work—except when its length exceeds about twice its diameter—requires no support from the back centre, so that it can be drilled or hollowed out, or the rest can be fixed either at right angles or obliquely to the axis, and the 'face' of the work be turned up in any manner that may be required. Besides metal cup-chucks, the turner frequently supplies himself with others made of hard wood, 'tapped' to fit the mandril screw. These he can turn away to fit any special or valuable piece of work; but in general a more ready means of effecting this purpose is to turn up a small piece of hard wood which has already been fixed in a metal cup-chuck, inasmuch as it is then effectually prevented from splitting. Special cases, however, will frequently be found to occur, in some of which wooden chucks are preferable; as, for instance, the 'spring cup-chucks' used for turning billiard and other balls. These consist of wooden cups slightly too small in diameter, having longitudinal saw cuts made in them, so that they spring open when the ball is pressed into them. These and similar temporary chucks, in the preparation of which some care is required, and which are likely to be of occasional future service, are within the true province of wood chucks. Another purpose for which they are much to be preferred is for work which can only be attached by being glued to the chuck, which is a frequent and very efficient way of fixing any pieces of wood, of either large or small diameter, which are to be turned 'plankways' of the grain. Where a hole in

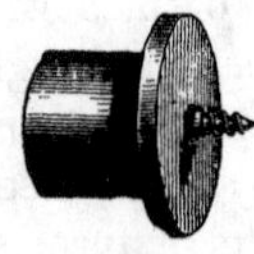

FIG. 111.
Screw-Chuck (¼).

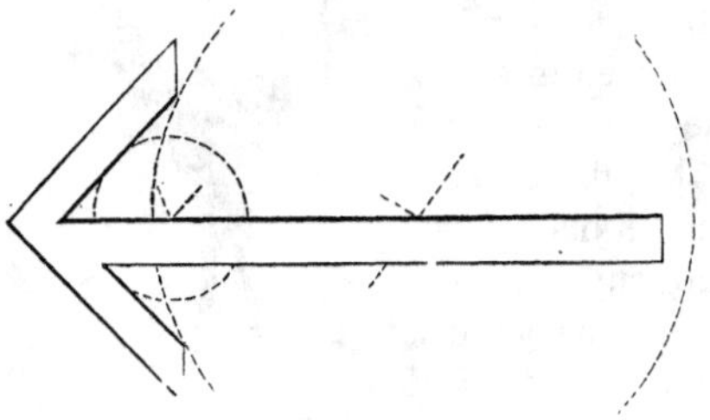

FIG. 112.—Centreing Square.

the centre of the back is immaterial, the *screw-chuck* (Fig. 111) can also be used in such cases.

In connection with this last chuck, we may notice the *centreing square* (fig. 112), a simple expedient for finding the centre of any circular piece of wood, or other material, of which the radius does not exceed the length of the blade. Bringing the V-shaped stock up to the circumference in any two positions, and passing a pencil or scriber along the blade, the point of intersection of the lines so drawn shows at once the centre of the circle. For metal and other work. of which the ends are circular and of small diameter, the centre can be still more readily found by means of the *centreing punch* (fig. 113), which is merely a pointed steel punch with parallel sides, sliding freely in the stem of an inverted funnel, or centreing cone. To whatever distance the circular end of the work may enter this cone, the point of the punch will be always at its centre, which spot can be marked by giving the top of the punch a light blow with a hammer. Without the cone, the pointed centre punch is to be met with in every workshop in which metal-turning or drilling is carried on.

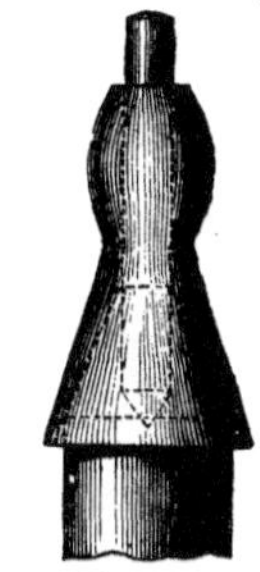

Fig. 113.
Centreing Punch.

Of chucks for the attachment of metal work to the foot-lathe, two kinds only demand attention here, namely, those used for the arrangement known as *Dog and Driver*, or *Driver and Carrier*, and the so-called American inventions, the *Scroll-chucks*. Most of the others are merely small sized copies of those in use with 'power-lathes,' in connection with which they will best be described. Indeed, the same might be said of the driver and carrier, but for large work an improved form of driver (Clement's) is now chiefly used, and we are not aware that it has at present been applied to foot-lathes.

The principle of the driver and carrier consists in supporting the work upon two steel points by means of a small centre hole at each of its ends, one point being attached to the mandril, the other to the poppet or back centre. It is thus rigidly supported, and is incapable of any motion except that of revolution. In order to cause it to revolve with the mandril, a dog or carrier is attached to the end of it, which lies towards the left hand of the workman, and a driver is so fixed to the mandril as to move with it. The forms of *Carriers* vary with the shape of the work to which they are to be applied, those represented in fig. 114 being amongst the most useful for circular and other sections. Fig. 115 shows the *Drivers* which are most frequently used in foot-lathes. The facility with which this system lends itself to work of the most varied shapes and sizes,—pro-

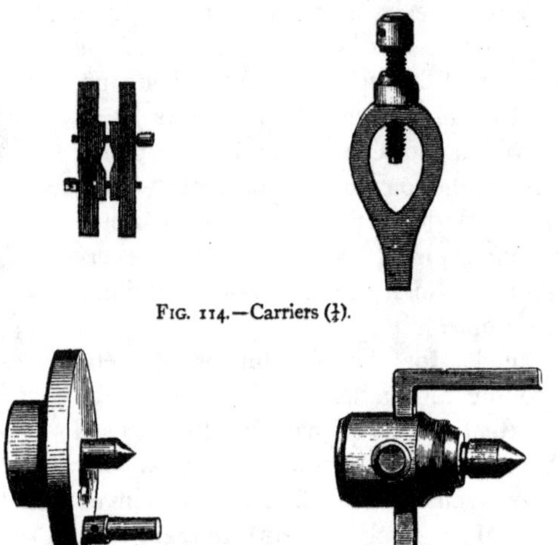

FIG. 114.—Carriers (¼).

FIG. 115.—Drivers (¼).

vided that its exterior only is to be turned,—together with the advantage which it presents in enabling it at any time to be again chucked concentrically, combine to make it of

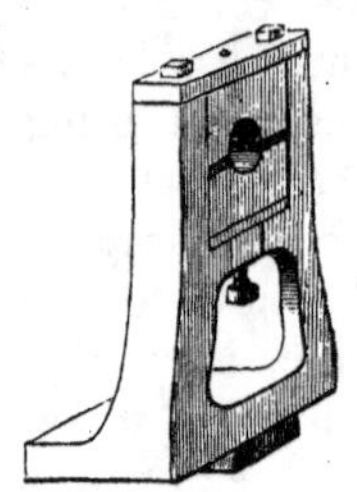

FIG. 116.—Metal Backstay.

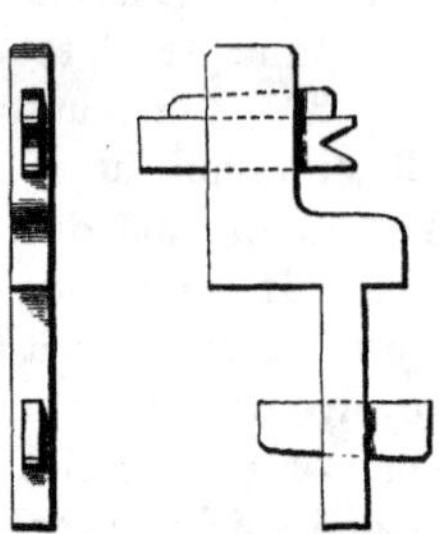

FIG. 117.—Wood Backstay.

very general application. The rigidity with which the work is thus supported—which is a very necessary quality when hard and unyielding materials are under treatment—is also much in its favour. Where, however, anything of considerable length and but small diameter has to be turned, the elasticity of the article itself is so great that some rigid support becomes necessary in addition to the two centre points. Such an apparatus is called a *Back Stay:* one form of it, adapted for supporting metal work of small diameter, being shown in Fig. 116, and one of a more simple kind, by which long and thin pieces of wood may be steadied, in Fig. 117.

For metal work which cannot be fixed between two centres, various other chucks are used, according to its size, shape, &c. Small pieces of no great length, especially of brass and the softer metals, are in general sufficiently rigid if merely driven into a piece of hard wood chucked in a cup-chuck.

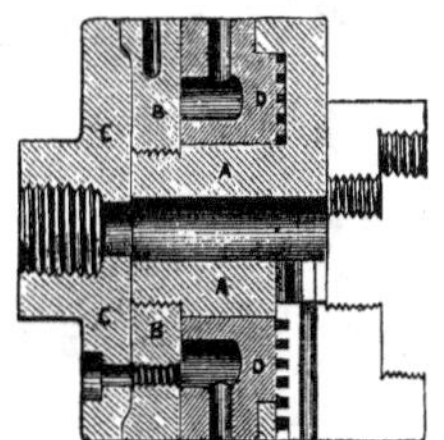

Fig. 118.—Scroll-Chuck, with Section through Centre (½).

'Bell-chucks' for pieces of large diameter and moderate length; 'cone-chucks' for discs having central holes, and 'face-plates' for flanged castings, and all kinds of work capable of being bolted down upon a flat perforated plate—are all made in miniature for use with foot-lathes; but the reader will sufficiently understand their construction and uses from the figures of their more powerful relations given in connection with 'power-lathes,' without special description of them here. For the 'four-jaw-chucks' there mentioned, an admirable substitute for small work has lately been introduced under the name of *American Scroll-chucks*, of one of which Fig. 118 shows a perspective view, together with a section through its centre. The great advantage which they possess is due to the jaws being all made to slide simultaneously, so that the necessity for 'centreing' the work held between them is entirely avoided. Every turner knows the sacrifice of time at which this operation has ordinarily to be effected, and will therefore be able to appreciate the value of any efficient means of dispensing with it. Its importance led to the adoption of a similar arrangement in Mr. Holtzapffel's workshop many years before 'American' self-centreing-chucks were heard of. The manner in which the motion of the jaws is made simultaneous will be understood from the section, where A, B, and C are three pieces which together form the main body of the chuck; when once put together, these to all intents and purposes form one piece. Enclosed between A and B is a ring, D, which can revolve independently. In the face of the chuck are three radial grooves, each of which has two feathers projecting from its sides into the body of the groove. Sliding freely along these grooves and feathers are the three jaws. On the face of the ring D a spiral of about $3\frac{1}{2}$ revolutions is cut, of such width and depth that its section resembles that of a square-threaded screw. Counterparts of the spiral are formed on the internal edge of each of the jaws, which, when in position, are in contact with the face of the ring D. Consequently, by causing this ring to revolve while the body of the chuck is at rest, the three jaws are all made to advance or to recede from the centre to an equal extent. This is readily effected by introducing two pins, or—what is better—two chuck-wrenches, of the form shown in Fig. 119, into the holes made for that purpose in the circumferences of B and D.

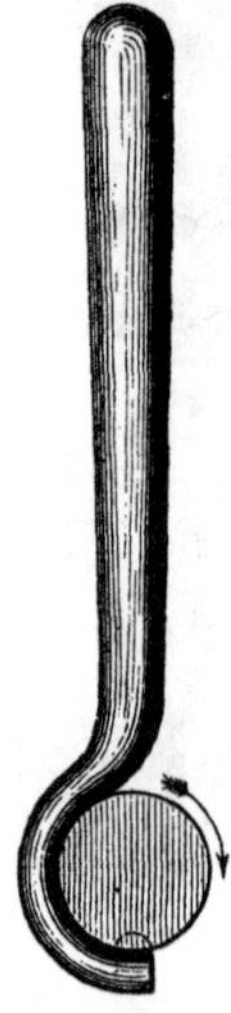

FIG. 119.
Chuck-Wrench.

Although the 'stepped' form of the jaws enables them to take hold of pieces of work of which the diameter varies considerably, these chucks cannot be employed for holding fine drills or similar small articles. For these, one of the various forms of chuck resembling that shown in Fig. 120 can hardly be too highly recommended. They are known as *Drill-Chucks*, and are much used in conjunction with the spiral drills figured in Chapter III. Like the preceding, they have three sliding jaws, which can be moved simultaneously in a similar manner, so that the drill, if straight to begin with, cannot fail to run true when fixed in the lathe.

Fig. 120.—Drill-Chuck (½).

Before the use of these chucks became general, metal work, when of small diameter, was frequently fixed in plain metal chucks, bored out almost to the same size as the work, and split by cutting them longitudinally with a saw; the requisite tightness being obtained by driving up a ring on the outside, which was slightly tapered for the purpose. But it is evident that each of these chucks could be used only for articles differing from one another but little in diameter. With the three-jaw-chucks, on the other hand, the diameters may vary within comparatively wide limits, so that no enlargement or reduction of the stems of a set of drills is necessary; which in itself is a great advantage, to say nothing of the more efficient manner in which they are held.

It is frequently required to drill a truly central hole in a piece of work which is already fixed in the lathe. The drill is then pressed up to the work by means of the poppet, being at the same time prevented from revolving by holding it with a hand vice or a drill-chuck, or in any other convenient way. No special apparatus is necessary except in cases in which the poppet is already employed in supporting one end of the work. Recourse must then be had to a substitute for it, by which access to the extremity of the work shall not be prevented. Such a substitute will be found in the *Boring Collar* (Fig. 121), of which the essential part is a thick plate containing a series of countersunk holes of different sizes. Any one of these can be brought into the line of the axis of the mandril, provision being also made for fixing the plate at any part of the lathe bed; the circular end of the work supported in the countersunk hole can then be drilled from the opposite side of the plate.

FIG. 121.—Boring Collar (1/6).

Although our notice of foot-lathe apparatus must necessarily be a hasty one, two other kinds of chuck demand a short description—the 'eccentric' and the 'oval' chuck. But, in connection with these, it must be borne in mind that for the former some means of fixing the tool is absolutely necessary, and for the latter it is very desirable; so that an explanation of the slide rest, which affords the means of securing and giving the requisite motion to the tool, ought perhaps to have preceded them.

The object of the *eccentric-chuck*, which is used almost exclusively for ornamental turning, is to cause the work to revolve round any point within a moderate distance from its true centre. Thus, supposing Fig. 122 to represent the outline of a piece of wood, chucked and turned up; in the natural course of things it revolves round its centre, A, and without being chucked afresh, its axis cannot be in any other position. But with the aid of an eccentric chuck it can be made to revolve round any point, B or C; and by a simple arrangement a complete series of points, such as C C′ C″, &c., all equidistant from the true centre, can in turn be brought into a line with the axis of the mandril. By causing a fixed tool to trace a circle round

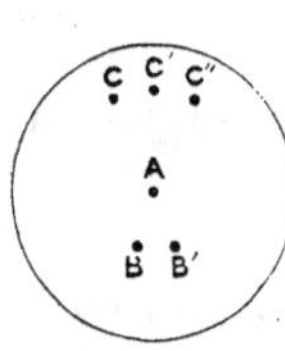

FIG. 122.

each of these points in succession, an elaborate and occasionally pretty species of ornament can be applied to the face of any piece of work. In Fig. 123 are two specimens of such patterns.

The requisite eccentricity is given in the following manner.

FIG. 123.—Eccentric Patterns.

Between the ordinary cup-chuck and the mandril there is introduced a thick brass plate (b, Fig. 124), carrying a second sliding plate (c) upon its surface. When the chuck is in the position shown in the figure, the V-shaped jaws, between which the plate (c) slides, are vertical, so that a portion of the ends of the plate only are seen. Inside a boss (a), which is cast in one piece with the first plate (b), there is a female screw for fixing it to the mandril, and a screw for the attachment of the cup-chuck containing the work is carried by the slide (c). By turning a handle applied to the end of its adjusting screw (d), a slow motion is given to the slide (c), the axis of the work attached to it being thus made more or less eccentric to the axis of the mandril. The amount of this eccentricity can be read off by means of main gradua-

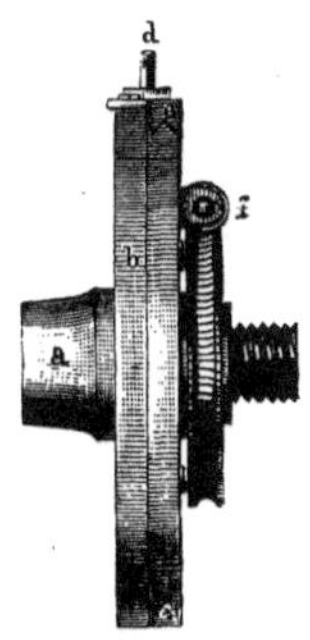

FIG. 124.
Eccentric-Chuck.

tions on the internal face of the plate and subdivisions on the head of the screw (d). To enable the successive points of a series (as c c′c″, &c., in Fig. 122) to become in turn the axes of rotation by being brought into a line with the axis of the mandril, the screw which carries the work is not rigidly fixed to the slide (c), but can be made to revolve slowly round its axis by means of the tangent screw (f), which takes into a wormwheel, the edge of which can be distinctly seen in the figure. In order that the circle may be readily divided into an equal number of parts, the teeth of the wormwheel are numbered, and the head of the tangent-screw is also graduated; but a simple notched wheel and catch is often used instead of the wormwheel shown in the engraving. In either case it should have some readily divisible number of teeth—the advantages and disadvantages of such numbers as 96 and 100 in this respect being a subject on which advocates of a decimal system would do well to ponder—100 being divisible only by 2, 4, 5, 10, 20, 25, and 50; whereas 96 is divisible by 2, 3, 4, 6, 8, 12, 16, 24, 32, and 48. It is therefore probable that the most constant lover of decimals would for once consent to abandon his favourite system, and, following the common practice, would have his eccentric chuck provided with 96 teeth.

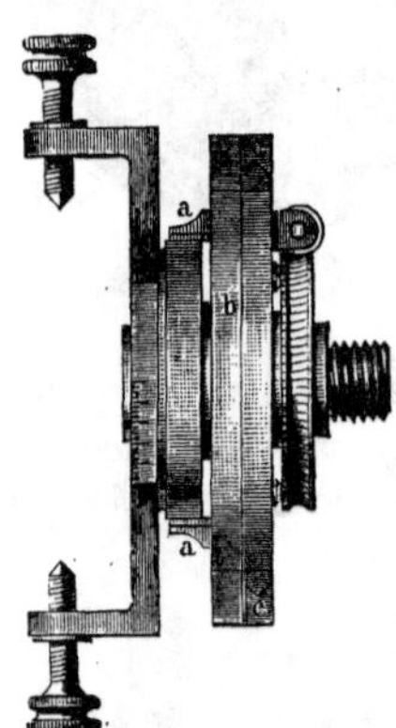

FIG. 125.—Oval-Chuck.

The *oval-chuck* almost requires to be seen to be understood, but a general notion of its appearance and action may be gained from Fig. 125. Its object is to impart such a motion to the work that a tool held stationary in contact with it shall at each revolution of the mandril describe an ellipse instead of a circle. The right-hand portion of the figure will be seen to bear much resemblance to the eccentric-chuck just described. A plate (b) has upon it a similar boss (scarcely seen in the figure), by which it is attached to the mandril; and a similar slide (c) has a screw for carrying an ordinary chuck containing the work, at the base of it being a wormwheel and tangent screw, as before. In passing, we may mention that these last are better omitted when the chuck is to be used for plain oval turning only, the screw being then part and parcel of the slide (c), so that one source of possible unsteadiness is avoided. Instead of the slide (c) being adjusted by the slow motion of a screw, as in the eccentric chuck, it is made to work in its groove with perfect freedom, carrying with it however two guides (aa), which are firmly fixed to it, and which pass through slots in the plate (b). Entirely detached from the above plate, slide, &c., is a ring, of which the outside diameter exactly fits the space between the guides (aa). By means of the arms and milled-headed screws on the left of the figure, this ring can be firmly fixed to the head-stock of the lathe without in any way interfering with the rotation of the mandril, the screw of which passes through the ring. As long, therefore, as this fixed ring is concentric with the mandril, the oval-chuck in no way influences the motion of any piece of work which may be attached to it, its guides (aa) merely sliding round the ring without being drawn out of their natural course. But when this ring is moved sideways by loosening one of the pointed screws and tightening the other, so that its centre no longer coincides with that of the mandril, it no longer allows the guides (aa), nor the slide (c), which is in connection with them, to follow a circular path. At two points, indeed, during each revolution of the mandril (when a line joining the centres of the guides is vertical, the arms and screws being horizontal when the ring is in position), the slide (c) is concentric with it; but after these points have been passed, the guides alternately draw the slide more and more from its natural course, till it has an amount of eccentricity equal to that at which the ring has been set. This occurs twice in each revolution; and, since the effects of these two disturbances are in opposite directions, the major axis of the ellipse thus described exceeds its minor axis by twice the amount of eccentricity given to the ring. By means of graduations on any convenient part of its arms, the amount can be read off. It must of course be equal to half the difference between the major and minor axis of the ellipse. Thus, for cutting an ellipse measuring $7\frac{1}{2} \times 6$ inches, the ring must have an eccentricity of $\frac{3}{4}$ of an inch. Some account of the theory of the oval chuck may be found in another volume of this series.[1] In using it with hand-tools and an ordinary T-rest, considerable care is necessary to keep the edge of the tool at one constant height above the lathe-bed. Any change in the height of the tool causes a corresponding change in the direction of the ellipse, accompanied by the destruction of the work, in the manner shown to an exaggerated extent in Fig. 126.

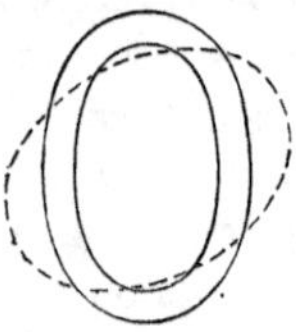

FIG. 126.

The *slide rest*, in its simplest form, consists merely of an arrangement for holding a turning-tool by mechanical means instead of by hand, and for giving to the operator, through the intervention of one or two simple adjustments, more complete control over it than he can otherwise obtain. Such an apparatus—which forms a most valuable addition to a foot-lathe—is represented in Fig. 127. On the under side of its base is a wide projecting feather (a), which fits into the space between the two cheeks of the lathe-bed; in this is a tapped hole, which enables it to be firmly bolted down upon the bed. On the base plate is a sliding plate (c), containing a dove-tailed groove, to which a steady backward or forward motion can be given by turning the handle (b), which works a horizontal screw. Attached to this plate (c)

[1] *The Elements of Mechanism*, by T. M. Goodeve, M.A., pp. 210–11.

by two vertical screws and a central pin is the lower half of the cross-slide (d), on which the upper portion of the rest (g) can be made to travel laterally by turning the handle (*f*) at the end of a second horizontal driving screw. These two slides together form what is sometimes termed a 'compound' slide. The tool being firmly clipped in the tool-holder (at the top of the figure) by screwing down upon it any two of the four screws, it is evident that a slow motion can be given to it, either parallel to the axis of the mandril by turning the handle (*f*), or at right angles to it by turning the handle (b). Or, by variously combining these motions, any

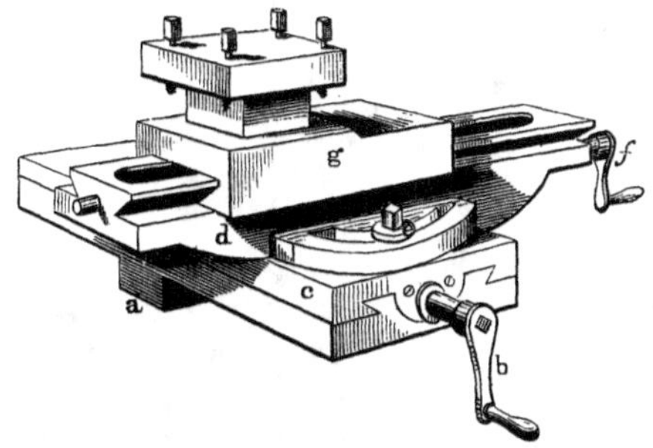

FIG. 127.—Slide-Rest for Foot-Lathe.

curved or conical surfaces can be turned—the latter, however, being more easily effected by placing the cross-slide in an oblique position and moving the tool by means of the handle (*f*) only.

Such is the general principle of the slide-rest, the introduction of which has been followed by the mighty results to which we have already alluded. Since its efficiency is either greatly impaired or altogether destroyed if the tool, when fixed in the tool-holder, possesses any 'shake' or unsteadiness, it is evident that the slides must be formed with the greatest care—the two halves of the dovetail being in every case kept perfectly parallel, and the surfaces which are in contact being made truly plane. Thus, and thus only, can we obtain evenness of fit at all parts of the slides, without which the uniform smoothness in their movement and steadiness on the part of the tool supported by them, which are the tests of the efficiency of the slide-rest, are impossible.

FIG. 128.—Spherical Slide-Rest.

But even as applied to foot-lathes, the slide-rest is by no means always equally simple.

Some means of adjusting the height of the tool is often a desideratum, and for ornamental turning a third slide is frequently interposed between the cross-slide and the tool-holder; occasionally, also, even a fourth slide and other adjustments are introduced—as in the 'spherical slide-rest' represented in Fig. 128—by which wonderful things may doubtless be effected. But with each fresh complication an additional chance of unsteadiness in the tool is introduced, which can only be avoided by increased care in the manufacture and use of the apparatus; so that, except for the purpose of producing elaborate patterns (which are generally of doubtful beauty), a slide-rest which is so constructed as to hold the tool as rigidly as possible is much to be preferred to one of great complexity.

In the slide-rest (Fig. 127), the tool-holder consists of a square plate, fixed upon the upper portion of the cross-slide (g), at a height somewhat exceeding the thickness of the tool, which is held by screwing down upon it any two of the screws at the corners of the plate. This arrangement only admits of the tool being placed either parallel to the direction in which the cross-slide moves, or at right angles to it, which is frequently a disadvantageous position for it. Other forms of tool-holder are therefore generally adopted. In one of these, which was proposed by Professor Willis in the year 1842, the upper plate is made triangular, a holding-down bolt passes loosely through its centre, and at its three angles are three bearings; two of these press evenly upon the tool when the screw which constitutes the third bearing is tightened. Consequently, half a turn of this screw suffices to fix or to liberate the tool; and it can be placed at any angle with the axis of the work. Another tool-holder, which affords similar facilities for fixing the tool at any angle by means of a single screw, and is rather more compact in appearance, is shown in Fig. 129. When it is not exerting pressure upon the tool, the slotted cylinder can be turned round on its axis. Its inability to hold any tool which cannot be passed through the slot is its chief disadvantage. But none of these tool-holders are now much employed

FIG. 129.
Circular Tool-Holder.

except in the case of the small slide-rests, of which the use is confined to the foot-lathe. In those of larger size, the only permanent provision for holding the tool is generally the introduction of four bolts into the upper portion of the cross-slide. These are fixed with their heads downwards, and the tool is secured by being placed beneath two short perforated bars, which are slipped over the ends of the bolts, nuts being then screwed down upon them. The arrangement may be observed in the engraving of the duplex lathe (Fig. 143), and elsewhere in the next chapter, two only of the holding-down bolts being in use, and the opposite

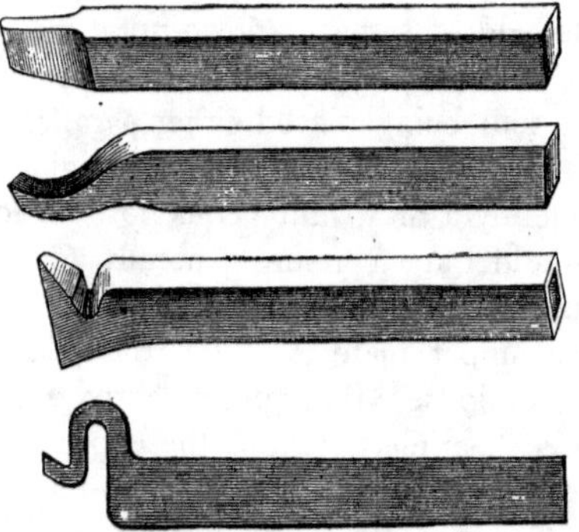

FIG. 130.—Slide-Rest Tools for Hand-Turning.

ends of the bars being supported by a short packing piece of the same height as the tool.

Examples of the slide-rest tools adapted for hand-turning are given in Fig. 130, the uppermost of them being known as a 'round nose,' and the others as a 'square nose,' a 'hook tool,' and a 'spring tool' respectively. But many which are used for small work in foot-lathes differ only in respect of size from those required for heavy work in power-lathes; and of these the principal typical forms will be found at a future page (see Fig. 150). Fig. 131 shows two kinds of *cutter-holder*—an excellent arrangement—the principle of which was first suggested by Mr. Babbage. Considerable economy results from their use. In the first place, a very much smaller quantity of tool steel is required, the holder being made once for all of steel of inferior quality; and, secondly, they are much more easily sharpened. Thus, the ordinary diamond-pointed hook tool shown above must be carefully forged and tempered to begin with; and when blunt, must be ground at the top and on its two sides. A similar tool can be made by fixing in an appropriate cutter-holder a short piece of steel of triangular section, which can be sharpened by grinding its upper face only. The necessity for adjusting the height of the rest as the tool gets worn away in sharpening, is also dispensed with when holders

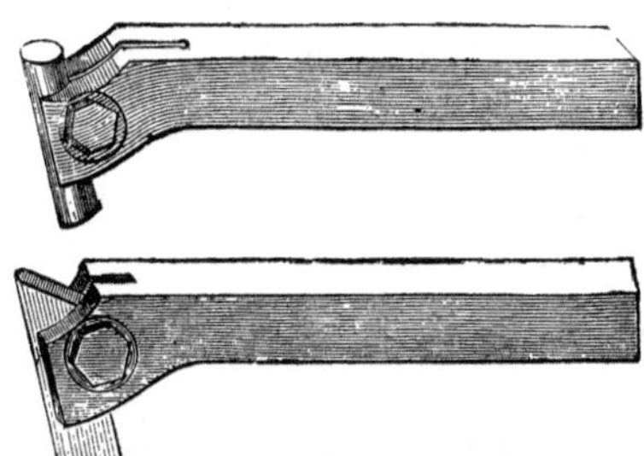

FIG. 131.—Cutter-Holders.

with loose cutters are substituted for the usual hook tools. Altogether their general adoption, both for light and heavy work, seems likely to be only a matter of time. Several patterns besides those shown in Fig. 131 are now in use, of one of which a representation will be found in Fig. 151.

For ornamental turning, slide-rest tools of much smaller dimensions than those above represented are used. Some of their forms are shown (full size) in Fig. 132, but they are made in very great variety to suit the wants or fancies of the amateur.

In many of these operations motion is imparted to the tool only, so that the process becomes more nearly allied to drilling than to turning. The mandril is then useless except for affording a firm and readily adjustable support for the work in hand, its graduations (on the divided plate already mentioned) enabling each point of any series to be brought in turn into a line with the axis of the revolving cutter. Easily divisible numbers—such as 96, 144, and 360—are chosen for the different 'scales' on the divided plate, for the same reason as for the teeth of the eccentric chuck. Lathes intended for this kind of work are provided with some kind of 'overhead motion,' for enabling the rotation of the flywheel to be communicated to the tool supported by the slide rest. This is generally done by one or other of the following arrangements. Either a light shaft is placed parallel to the axis of the mandril at some three feet above it, and this is driven directly from the fly-wheel without

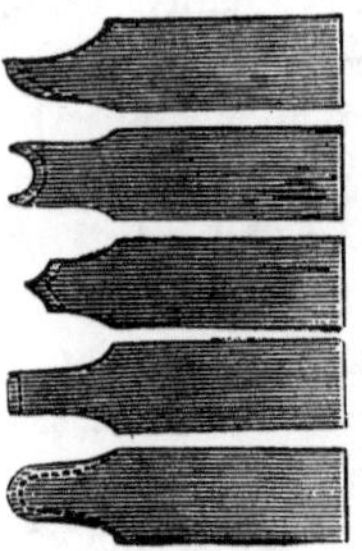

FIG. 132.—Ornamental Slide-rest Tools.

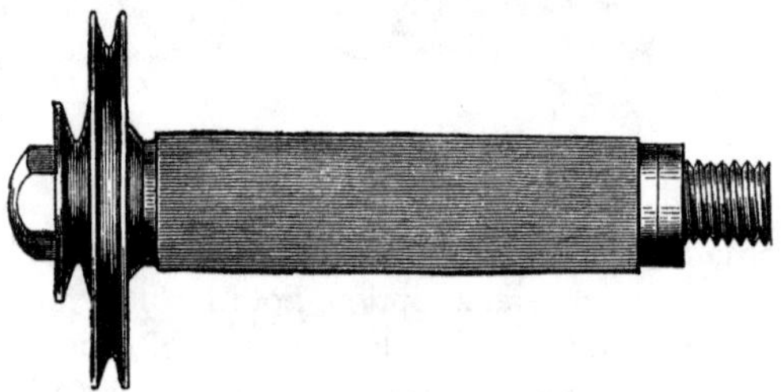

FIG. 133.—Revolving Cutter-Holder.

being in any way connected with the mandril; a second short band being then passed over a pulley on this shaft, and round the revolving cutter-holder. Or a long single band is used, which runs first round the flywheel, thence over two loose sheaves supported by a balanced bar above the head of the operator, and which finally descends, as before, to the revolving holder in which the cutter is fixed. Fig. 133 represents a *Revolving Cutter-Holder*, which can be used in this way for either plain or slot drilling, for cutting grooves, key beds, &c., and for a variety of ornamental purposes. It consists of a square shank, of a convenient length for being held in the tool-holder of the slide rest, within which is a steel spindle, having at one end a grooved pulley, and at the other end a screw, and also a central hole for the attachment of the drills or cutters. These having various forms of cutting edge (examples of which are given in Fig. 134) produce by their revolutions a great variety of beaded and other patterns. Or, by substituting for these round or flat-sided drills, and giving them a horizontal motion by means of the slide rest, grooves of corresponding sections can be cut either parallel or transversely to the axis of the work.

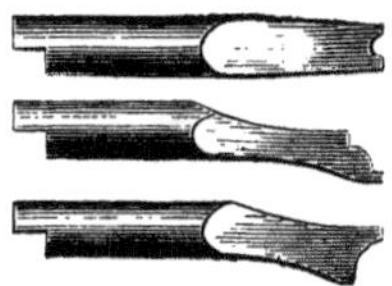

FIG. 134.
Revolving Cutters for Ornamental Turning.

Extending the system of imparting the motion to the tool instead of to the work, a slight addition to the revolving holder just described enables us to dispense altogether with the eccentric chuck for engraving such patterns as those given in Fig. 123. For this purpose the cutter, instead of being attached concentrically to the spindle, is fixed in a holder which can be made to travel along a cross-slide; as shown in Fig. 135, where a is the square shank, b the grooved pulley at the end of the spindle, which revolves within it, and c the cross-slide. The holder can be adjusted by means of a milled-headed screw and can be firmly clamped at any part of the slide, graduations on the head of the screw enabling it to be accurately set, to as to describe any circle whose radius does not exceed the length of the slide. Series of intersecting circles can then be traced on the face of the work as before described, their positions being arranged with the assistance of the divided plate of the mandril instead of the eccentric chuck.

By an ingenious, but slightly complicated arrangement of

cutter-holder, the oval chuck also may be dispensed with in similar cases, the cutter being made to travel in an elliptical path which may vary to any required extent,

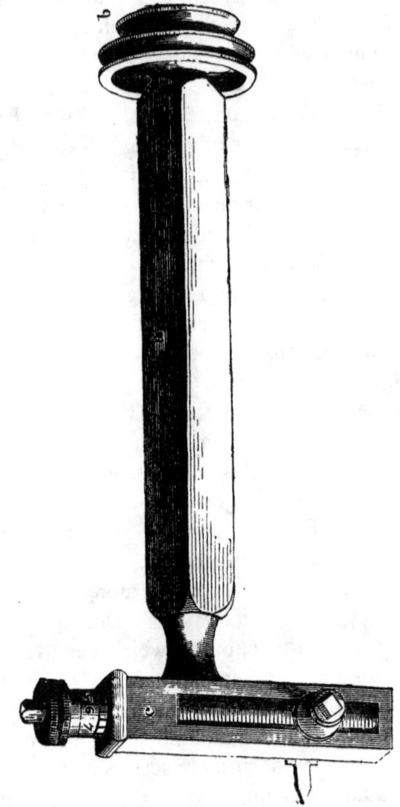

FIG. 135.—Eccentric Cutter.

from a circle on the one hand to a straight line on the other. Other and yet more elaborate forms of apparatus, and also the combination of some of the above with one another and with the motion of the mandril, result in the

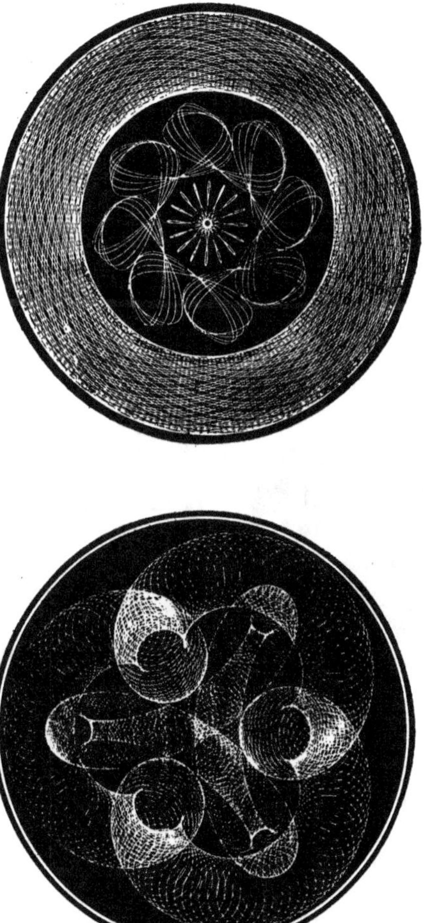

Fig. 136.—Geometric Chuck Patterns.

production of face-work patterns, &c., of great complexity and occasional beauty. In Fig. 136 are two examples of such patterns, which will serve to show the capability of the *geometric chuck;* a most ingenious contrivance by which an endless variety of graceful curves can be obtained. For these and for many of the previous figures in this chapter, as well as for two of the subsequent ones, we are indebted to Mr. Northcott, in whose work on the subject of turning[1] a drawing and description of this chuck may be found, together with various devices engraved by it. To that work we must also refer the reader for an account of an excellent substitute for the *Rose Engine,* by which such patterns as Fig. 137 are produced. The cases of watches, to which, on account of their curvature, the ordinary eccentric patterns are inapplicable, are generally ornamented in this way, the waviness being formerly produced by attaching to the mandril a circular plate with the requisite number of notches round its edge. Against this a small roller was made to bear, which, entering each of the notches in turn, caused a slight corresponding lateral motion on the part of the mandril, the headstock being so arranged as to admit of it. Thus the work was made to vibrate or 'chatter,' so that a fixed tool produced upon it a pattern with the same number of waves as the plate or 'rosette.' In the 'rose-cutting instrument' referred to, the cutter is made to vibrate, the work being stationary; an arrangement which has obvious advantages.

FIG. 137.—Rose Engine Pattern.

One other process peculiar to the foot-lathe deserves a

[1] *On Lathes and Turning*, Northcott. Longmans: 1868.

passing notice—that of cutting screws by means of a *traversing mandril.* For this purpose the two bearings of the mandril are made cylindrical, instead of being conical or of the form shown in Fig. 99, so that it can travel longitudinally for a short distance without its fit in the headstock being impaired. A short length of screw of the exact pitch required being made fast to the end of the mandril, a corresponding female screw is fixed to the headstock in contact with it. The mandril, when made to revolve, thus receives a slow and uniform longitudinal motion, and an exact copy of the screw can in this way be cut by a stationary tool firmly held in the slide rest, or even in the hand.

Two other kinds of revolving tools should also be mentioned, viz. :—*Milling Tools and 'Circular Cutters.'* The former are used for giving the requisite roughness to the heads of thumb screws (such as those in Fig. 125); and they have the merit of being very quick in their operations, a few turns of the lathe only being required in order to convert a plain circular moulding into a ribbed or beaded one. The tool consists of a small disc of steel, having all round its edge a counterpart of the pattern which it is intended to produce. This disc, which is mounted in a handle so that it can rotate with perfect freedom, is merely pressed against the revolving work with sufficient force to cause it to leave its impression. The discs are generally of small size—rarely exceeding half an inch in diameter.

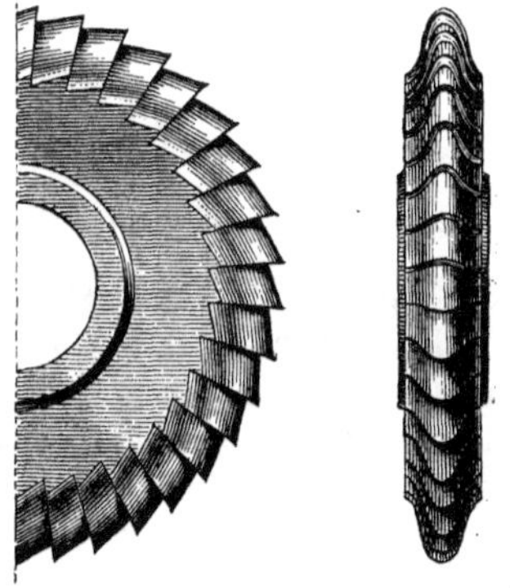

Fig. 138.—Circular-Cutter.

Circular Cutters are also made from discs of steel. In general they are of much larger diameter than milling wheels, to which in fact they bear no resemblance, since they perform their work by a true cutting process. They are used for forming the teeth of wheels and for various other purposes, being mounted, when in use, upon a spindle which can be driven from the flywheel and brought up to the work by being attached to the slide-rest. In Fig. 138 one of these cutters —or rather a portion of one—is represented full size. It will be observed that its teeth resemble those of a single cut file, and that during its revolution they act continuously in a precisely similar manner. The method of using these cutters, as well as the description of a revolving-cutter machine for making them, will be found in Mr. Northcott's work, already referred to.

CHAPTER VI.

ON POWER-LATHES.

THE lathes which we have thus far been considering are merely mechanical means of applying muscular power to the various processes comprised under the head of 'turning.' This, as we have seen, is generally effected by means of a treadle, crank, and fly-wheel, although a crank worked by hand, driving a fly-wheel of large diameter, is occasionally substituted. But although in this way some additional power may be applied, it would be utterly insufficient for the bulk of the work which has to be turned in an engineer's workshop at the present day, for which *power lathes* of various kinds have become an absolute necessity. Under this head we shall include all lathes which can be worked only by being connected with a steam engine, or some other independent machine.

The transition from foot-lathes to power-lathes might seem

indeed as if it ought to be accompanied by a diminution in the number of their working parts, although of course they must be of greater strength and weight; but a glance at the engravings in the present chapter will soon dispel any such idea—the cause of their apparent complication being, broadly speaking, twofold. In the first place, most of the work which is performed with them requires a very slow speed; and for the transmission of this the leather belts by which the power is, as it were, laid on from the engine, or other prime mover, are not well adapted. Consequently, their comparatively high velocity must be much reduced before it is communicated to the work, and although this is in almost all cases partially effected by means of a 'counter-shaft,' arrangements for still farther exchanging speed for power are affixed to the headstock of every power-lathe.

Somewhat greater complications, however, are rendered necessary by the universally prevalent practice of making power lathes 'self-acting'—that is to say, providing them with some mechanical means of giving the proper feeding motion to the tool. This may be effected through the agency of the slide-rest in several different ways, one of the most frequent being that adopted in screw-cutting lathes of connecting it with a long and accurately pitched screw (termed the 'leading screw') of which the axis is parallel to the lathe-bed. For this purpose the driving speed must be still farther reduced, and the proportion which the revolutions of the leading screw bear to that of the mandril must be definite and capable of adjustment; conditions which are most easily and certainly fulfilled by connecting them together by a train of 'change wheels.' These, especially when a lathe is also provided with the 'self-acting rack-traversing motion,' hereafter described, give at first sight an appearance of great complexity. But this is soon dispelled when each set of parts is examined independently of those which, although very near neighbours, are in no way related to it. Each, however, of the above self-acting arrangements

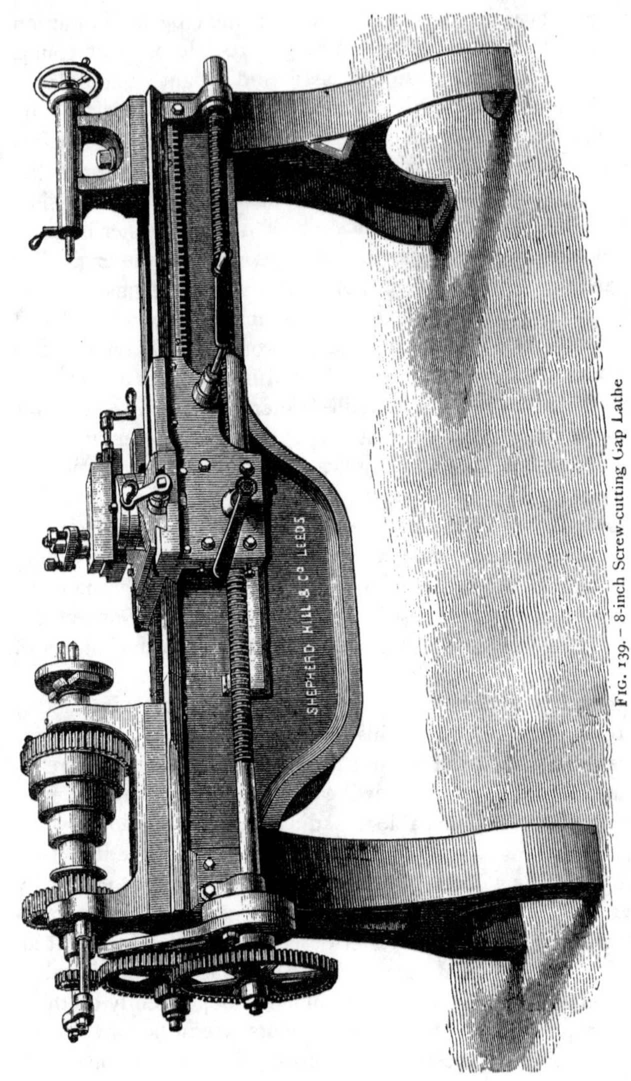

Fig. 139.—8-inch Screw-cutting Gap Lathe

requires corresponding additions to be made to the slide-rest, which consequently assumes proportions very different from those of its simple ancestor shown in Fig. 127. But it will be better to defer the detailed consideration of the various parts of a self-acting slide-rest till the description of one or two of the lathes themselves have rendered evident the necessity for these modifications in its construction.

Fig. 139 shows an 8-inch *screw-cutting gap lathe*, which will serve to explain the double-geared headstock by which the requisite slowness of motion is given to the mandril, and also to show the position, &c., of the leading screw, and the change wheels by which such lathes are rendered self-acting. With the difference between a plain and a 'gap' lathe, we will not trouble ourselves at present. First, then, as to the *double geared headstock*, of which a plan is given in Fig. 141, the lower part of the drawing being in section so as to show the form of the mandril, which bears but little resemblance to that of the foot-lathe given in Fig. 99. Both its bearings in the headstock are conical, the front or right hand cone being formed on the mandril itself, the hinder or left hand one being a loose conical bush carefully fitted to it. This, among other advantages, enables an accurate fit to be maintained between the mandril and the headstock, without having recourse to plummer-blocks, which would be necessary if the bearings were cylindrical. (It may be observed that the headstock in Fig. 98 has plummer-blocks on this account.) On the mandril—of which the whole of the central part is of uniform diameter—is a loose, conical pulley, to the small (hinder) end of which a pinion is attached. These are in no way connected with the mandril, but revolve freely upon it. Fig. 140 will explain the arrangement, F and G being the conical pulley and pinion, inseparable from each other. but not connected with the mandril. In front of the large end of F is the spur-wheel L, which is connected with the mandril, and always revolves with it; and gearing into this is a pinion K, keyed upon a short shaft parallel to the mandril. Fixed upon this shaft there is also a spur-wheel H, gearing into the pinion G. The effect of driving the pulley F at a quick speed by means of a leather belt (of which a portion appears in the diagram) is thus to make H revolve less rapidly; the pinion K, which runs at the same speed as H, still further reducing it in its transference to the mandril through the spur wheel L. To enable the mandril to be driven at a quick speed when required, a bolt can be screwed through L into the face of the pulley F, thereby connecting it with L,

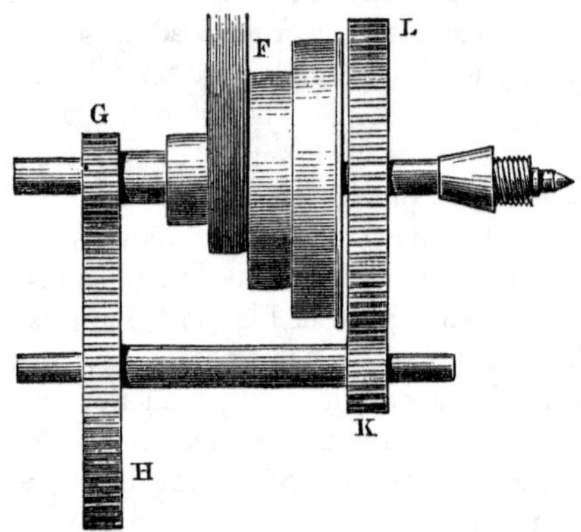

FIG. 140.

and through L with the mandril. H and K must then be thrown out of gear, which can be done either by sliding the shaft which carries them in the direction of its length (as described and figured at p. 142 of Professor Goodeve's volume of this series), or by the somewhat neater expedient of supporting it in eccentric bearings, instead of directly in the headstock itself. The headstock represented in Fig. 141 is fitted with this arrangement, the portions of the framing which would naturally conceal the discs (B and B′) being removed; as also are parts of the pulleys, &c., for the purpose of showing the mandril, as already explained. One of these discs (B) is likewise shown in section, so as to exhibit the position of the shaft C, when the toothed wheels which it carries are in gear with those on the mandril. When the wheels are out of gear, the shaft occupies the position indicated by the dotted lines in the sketch (Fig. 142), in which C represents the end of the shaft under the former

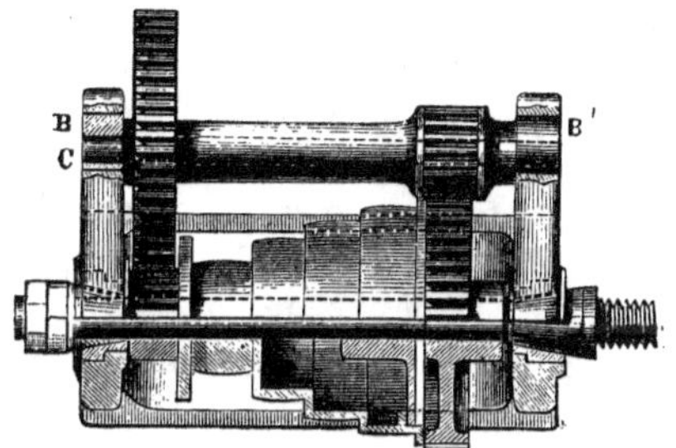

FIG. 141.—Plan of Headstock.

condition, and B the disc in which it is supported eccentrically, of which the 'throw' need only exceed the depth of the teeth of the wheels by a sufficient amount to ensure their clearance. It will be observed that this system, which is now very largely used, not only for lathes, but also for drilling machines, &c., admits of the headstocks being somewhat shorter than is required for the sliding arrangement alluded to above.

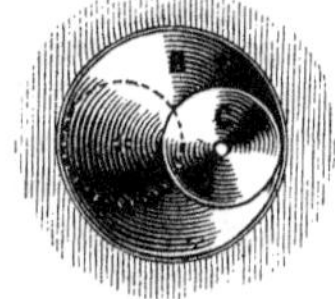

FIG. 142.

In a plain 'screw cutting' lathe,—and it was for the purpose of originating screws that Mr. Henry Maudslay, at the commencement of the present century, devised the self-acting lathe,—a shaft with an accurate *leading screw* cut upon it is supported (in bearings at its ends only) in some convenient position truly parallel to the axis of the mandril. In Fig. 139 it is placed in front of the lathebed. In Fig. 143 it is between the cheeks of the bed and therefore cannot be seen. In either case the connection between the slide rest and the leading screw can be made or broken at pleasure, and any requisite proportion between its speed and that of the mandril can be obtained by properly combining the *change wheels;* a set of which (consisting of about 1½ or 2 dozen) is provided with every lathe of this kind. The exact pitch of the leading screw being known (it has generally either 1, 2 or 4 threads to the inch)—a simple calculation shows what wheels must be employed for cutting a screw with the same or any other number of threads to the inch. For ordinary pitches, however, a table[1] is provided with each lathe, which shows at a glance the wheels which will give the proper result. But the correctness of the table, or the increase or diminution of speed—and by consequence the theoretical loss or gain of power—due to any other train of wheels (as, for example, the wheels of the double-geared headstock just described) is so easily calculated, involving only a knowledge of simple arithmetic, that all who study or use such lathes should be familiar with the process.

A clear explanation of it, and also of the method of so arranging the wheels as to cut left-handed screws, together with a diagram illustrating the action of a similar train of wheels to that shown in Fig. 139, will be found at pp. 136 to 139 of Professor Goodeve's 'Elements of Mechanism.'

The actual arrangement of a set of change wheels can be clearly seen in Fig. 139. A spur-wheel is attached to the end of the leading screw, and a pinion to the hinder end of the mandril, which is prolonged beyond the left hand bearing of the headstock for that purpose. This projecting end of the mandril is supported by two horizontal pillars carrying a cross piece, the latter serving to assist the foremost conical bearing in taking any end thrust which may be put upon the mandril. Connection is established between the aforesaid pinion and spur-wheel either by a single intermediate-wheel or by an intermediate-wheel with a pinion attached

[1] See Table of Change Wheels, at end of this volume, for Lathe, Fig. 139.

to it;—the latter of these arrangements being shown in the engraving. But the variable sizes of the pinion, spur, and intermediate-wheels required for driving the leading-screw at different speeds render it necessary to give the intermediate-wheels an adjustable support; for which the universal practice is to mount them on a *Radial Arm* or *Tangent-Plate* instead of on a fixed bearing. This arm is moveable about the axis of the leading-screw, and by means of set-screws working in circular slots it can be clamped at such an angle as to enable the intermediate-wheel to gear into the pinion on the mandril. At the same time either this wheel itself or the pinion attached to it can be brought into gear with the spur-wheel of the leading-screw, provision being made for variation in their diameters by grooving or slotting the radial arm, and thereby enabling the intermediate-wheel spindle to be fixed at any point throughout its length.

Running along the front of the lathe bed above the leading screw, in Fig. 139, a rack may be observed. Its object is to enable the slide-rest to be moved along the lathe bed (or 'traversed') by hand when it is not required for screw-cutting, and also in screw-cutting to enable the rest to be run back quickly before commencing a fresh cut. A pinion (not seen in the figure) gears into this rack. In some cases the screw itself is made to serve for a rack, a worm wheel being substituted for a pinion; but as so much depends upon the accuracy of the leading-screw, all unnecessary wear and risk of injury to it should be avoided.

The construction of the *back centre* (sometimes called the loose headstock) will be sufficiently intelligible from the section of the foot-lathe poppet head previously given (in fig. 99). Some description of the chucks used with this and other power-lathes will be found at the end of this chapter.

Fig. 143 shows a larger lathe of somewhat different construction from the preceding one. The bed is parallel throughout its length, and the leading screw, as has already been pointed out, is placed between the cheeks of the bed, so that it cannot be seen in the engraving. Where a duplex slide rest is used—an invention of Sir J. Whitworth's, by the adoption of which much time may be saved—this central position of the leading screw has great advantages. The double toolholders can be clearly seen in the figure; their action will be explained when we come to consider the self-acting slide-rest generally.

In this, however, as well as in the lathe shown in Fig. 139, the slide-rest depends for its self-acting motions upon the leading-screw only; and when this is the case, its original accuracy (as previously stated) cannot long be maintained. This is to some extent due to the greatly increased strains to which it is subjected when used for turning heavy work instead of for cutting the threads of screws only. But, even if every possible care be taken in this respect, a still more fertile source of error remains, from the fact that the bulk of the work in almost every lathe is performed within a short distance of the mandril, so that this part of the screw becomes much more rapidly worn and deteriorated than the remainder. This has led to the introduction of self-acting apparatus for giving motion to the slide-rest by means of the longitudinal rack, to which, in its application to hand-traversing, attention has already been directed. The gearing required for this purpose gives an additional appearance of complication to lathes which are provided with it. These are known as *self-acting, surfacing, and rack-traversing* lathes; their slide-rests being capable of being driven either with a 'traversing' motion parallel to the lathe-bed, at right angles to it for 'surfacing,' or diagonally by a combination of the two. For this purpose a longitudinal *back shaft* is placed behind the bed, being supported (at its ends only) in the same manner and in much the same position at the back, as that in which the leading-screw shown in fig. 139 is supported in front. But this shaft, instead of having a screw-thread cut directly upon it, carries a short worm which, whilst it can slide freely in the

Fig 143.—10-inch Duplex Lathe.

direction of its length, is compelled always to revolve with it, the shaft being provided with a longitudinal groove, and the worm with a feather, for this purpose. In this way any part of the backshaft can be made to serve as a worm or tangent-screw to a worm-wheel which is carried by the slide-rest, whilst at the same time the rest can be made to traverse by hand, for which special arrangements would be

FIG. 144.—Back view of Rack Traversing Lathe.

required if the screw were continuous. The manner in which the various motions are communicated through the worm-wheel to the turning-tool will be fully explained when we deal with the slide-rest; at present, we will confine ourselves to describing the ordinary method of driving the backshaft, and of varying its speed, the arrangements for which are so clearly shown in the accompanying engraving that our task will be a comparatively easy one. This (fig. 144) represents a back view of the headstock end of a rack-traversing and surfacing lathe, made by Messrs. Shepherd, Hill & Co., and in it the back shaft worm and worm-wheel just mentioned can be plainly seen. It is also evident that the slide-rest, which carries the worm-wheel and also the toothed wheel and pinion in front of it—and which in the engraving is shown almost as close to the headstock as it is possible for it to be—cannot be moved along the bed without carrying the worm upon the backshaft with it. This, therefore, bridges over the main difficulty in transferring a portion of the power from the mandril to the tool, namely, that arising from the variable position of the slide-rest.

Tracing the course of the motion from the worm and shaft towards the mandril,—although it is of course not transmitted in this direction but in the opposite one,—we observe in the first place a spur-wheel made fast upon the extremity of the back shaft. Into this gears one or other of a pair of smaller spur wheels, according to the direction in which the back-shaft has to be driven. These wheels are supported not directly upon the end standard, but upon a rocking-frame, which can be moved backwards and forwards through a short distance by means of a bar terminating in a handle at the front of the lathe. Upon the position of this frame, as we have said, the direction of the rotation of the back-shaft depends. For the two wheels which it carries are constantly in gear with one another, one of them being also constantly in gear with a pinion attached to the conical pulley which appears below these toothed wheels in the engraving. When therefore the frame is in one position, the wheel which is always in gear with the pinion is brought also into gear with the backshaft-wheel, the second wheel running idle, but when the position is reversed, this second wheel is brought into the train, and the movement of the shaft is consequently reversed.

From the upper conical pulley to the lower one a belt transmits the motion, the three 'steps' in each affording the means of driving the back-shaft at three different rates of speed, whilst a spur wheel carried by the radial arm establishes the connection between the upper pulley and the mandril; the pinion on the overhanging end of the latter, by which it will be remembered that the requisite motion is obtained from it for driving the leading screw in screw-cutting, serving the like purpose, when not so required, in the case of the traversing and surfacing gear.

With reference to the use of the 'gap,' already mentioned, it will be rememberd that the chief limit to the capacity of a foot lathe was stated to be the height of the 'centre' (of the mandril) above the bed. A gap is an expedient for extending this limit, and thus enabling a lathe to take in articles of much greater diameter (if of moderate length) without materially increasing its weight or general dimensions. For this purpose the bed is cast with a gap just in front of the end of the mandril, where a corresponding bend may be observed on its under side in Fig. 139. When the gap is not required, a loose piece fitting closely into the opening in the cheeks gives the lathe an ordinary flush bed. In the figure referred to this piece is shown in position, and its fit is so accurate that the gap can hardly be recognised. Although a gap is undoubtedly useful, it is not altogether an unmixed advantage, inasmuch as it offers continual temptation to the workman to overtask the power of the working parts of a lathe, by enabling him to use it for heavier work than is advisable. The bed itself can of course be strengthened to any required extent by properly disposing an increased weight of metal, and in large gap-lathes the projection below the bed not unfrequently extends quite down to the floor of the shop.

But when the treatment of work of large diameter is the regular and not the exceptional duty of a lathe, the increase in size should not be confined merely to the distance between the mandril and the bed, but should be extended to all its working parts. As moreover it may often be necessary to turn up articles of which the length as well as the diameter is considerable, it is desirable to have the means of extending the limit which is set by the width of the opening in the bed of a gap-lathe. These requirements are fulfilled by what is known as a *Sliding Break Lathe*, of which an example is given in Fig. 145. In this the bed consists of two separate parts; a base-plate—which is of the full length of the machine, and of which the flat upper surface is kept at a sufficient depth below the mandril to enable the largest piece of work which the lathe can contain to run clear of it; and an adjustable upper part, or bed proper, which carries the slide rest and moving headstock upon its surface after the manner of an ordinary lathe bed, but of which the length is comparatively small. The width of the base-plate shown in our engraving is greatly increased in the neighbourhood of the fast headstock, to enable the slide rest, when mounted on a separate stand, to be used at a greater distance from the axis of the mandril than is possible when it is supported by the sliding bed,—but in some break lathes this arrangement is dispensed with. The width of the 'break' or 'gap' can be adjusted to suit the length of the work by moving the sliding bed bodily along the surface of the base plate; the back centre remaining always truly in a line with the axis of mandril. This traversing motion of course puts fresh difficulties in the way of rendering the slide-rest self acting. These however have been overcome by various expedients, consideration of the details of which would not be profitable in such a work as the present, so that we can only recommend the reader who is interested in them to examine the machines themselves when he has the opportunity. The increased number of toothed wheels at the back of the headstock in the engraving—which certainly at first sight gives a somewhat complicated appearance to this lathe—is due in part to the length of the train by which the necessary reduction of speed for the feed motion is obtained, and partly to the fact of the headstock itself being 'treble geared,' the face-plate being driven, when the work is of large diameter,

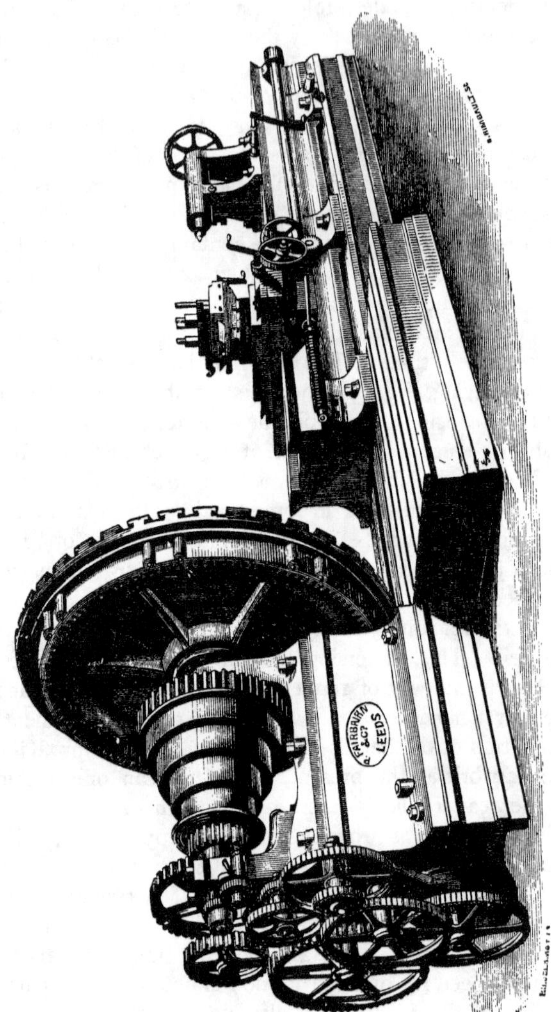

FIG. 145.—Break Lathe.

by a pinion gearing with the annular wheel at its back instead of by the mandril. These lathes are occasionally of very large size, the dimensions which determine their powers being, first, the height of the headstocks; secondly, the length of the bed; and thirdly, the diameter and length of the work which can be taken by the gap or break. In the figure the height of the headstocks (above the upper bed) is 21 inches, but this lathe is by no means a large specimen of its class; others ranging as high as 48 inches in the height of the head-stocks, with 20 feet beds and breaks capable of receiving work 14 feet diameter and 3 feet 6 inches wide.

When, however, articles which have to be turned are of such large size as regards their diameters, they are, more often than not, of but small width, in which case they can almost always be securely fixed without the assistance of the back centre. For these therefore the sliding bed may be entirely dispensed with, and the lathe may be greatly simplified and reduced in length. A small lathe of this kind—which is known as a *face lathe*, or *surfacing lathe*, from its principal duty being that of 'surfacing,' rather than 'traversing,'—is represented in the accompanying engraving (Fig. 146). The slide-rest, which is of a very simple form, is provided with a base of sufficient height to bring the tool to a level with the axis of the mandril. It is made self-acting (in this case in the direction for 'surfacing' only), by establishing overhead communication between the hinder end of the mandril and the cross-slide of the rest; a light chain being passed from the weighted arm at the front of the rest, over two sheaves, one of which is fixed above the slide-rest, and the other above the horizontal arm at the hinder end of the headstock. This arm receives an oscillating motion from a cam attached to the overhanging end of the mandril, which it thus imparts to the weighted arm of the slide-rest; from which it is transferred, through the pawl and ratchet-wheel at the end of the horizontal screw, to the tool contained in the tool-holder. The feed thus obtained is of course intermittent, but it can be varied considerably by altering the points of attachment between the extremities of the chain and the arms, which are slotted for this purpose.

When intended for surfacing work of very large diameter, it is advantageous for the speed of lathes of this kind to be duly proportioned to the distance from the axis at which the tool may be working, and they are sometimes provided with self-acting apparatus by which this proportion is maintained. If the speed be not varied, it is obvious that the duty per-

FIG. 146.—Face Lathe.

formed by both the lathe and the tool is much more severe when the cut is being taken from the outer extremity of the work, than when the central portions are being operated upon. So that in face-lathes of any considerable size, treble-gearing, or the means of driving the face-plate by a large toothed wheel attached to its back, such as that seen in the break-lathe (Fig. 145), becomes an absolute necessity. The above-mentioned capabilities of break-lathes indeed sink into insignificance, when compared with those which face-lathes occasionally possess. Machines of this kind with 14-foot face-plates, capable of turning articles up to 20 feet in diameter, may be obtained from any manufacturer of engineers' tools, and in bygone years the present writer was familiar with a face-lathe capable of taking in work up to a diameter of 40 feet. But this was intended for a special purpose—that of operating upon railway turn-tables (which, however, it has since become customary to turn when in a horizontal instead of in a vertical position)—and for special purposes we have at present arrived at no limits in respect of size which would be recognised by our machine makers as fixed and impassable.

Into the construction of the numerous forms of lathe which are intended for special purposes, and not for general turning, it will be impossible for us to enter here, although they would be found to be by no means wanting either in variety or in interest. But the objection which applies to detailed descriptions of machinery of every class—namely, that the constant introduction of improvements in its construction soon renders the description inaccurate—is of special force in their case, since almost every alteration in the size or form of the one article which each is designed to produce, renders corresponding changes desirable in the arrangement of the machine itself.

Turning now to the particulars of the self-acting slide-rest, a glance at Fig. 147 will show that as applied to screw cutting, rack traversing, and surfacing lathes, it is by no means the simple apparatus which was described in connection with foot-lathes. From our engraving its general appearance when in position upon the bed of a lathe of this kind may be seen, the front of the bed being provided with a leading screw and rack, and the back of it with the grooved shaft and worm, to all of which attention has already been directed.

First, with regard to the means of establishing the connection between the rest and the leading-screw—connection which must of course be capable of being made and broken at pleasure. This was formerly done by clasping the screw between the two halves of a horizontally divided nut which was attached to the slide-rest, and of which the upper and lower portions could be brought together or separated by

FIG. 147.—Self-acting Slide-rest.

raising or depressing a lever handle in front of the rest. The lever handle may be observed in the above engraving projecting below the bed of the lathe towards the left hand side of the rest; but it now acts upon a divided nut of increased length, of which one half (the lower) is entirely omitted. The mode in which it raises or depresses the upper half will be intelligible from the side view of the rest (Fig. 149); for which, and also for the plan (Fig. 148)—without which it would have been almost impossible to make this ingenious piece of mechanism clear—we are indebted to Messrs. Shepherd, Hill & Co., of Leeds. Careful study of the different parts in these three figures, in conjunction with the back view of the rest (Fig. 144) will indeed almost render verbal description unnecessary; and some farther assistance may be derived from a comparison of the present slide-rest with that which is shown upon the bed of the screw-cutting lathe (Fig. 139), of which the construction is more simple,

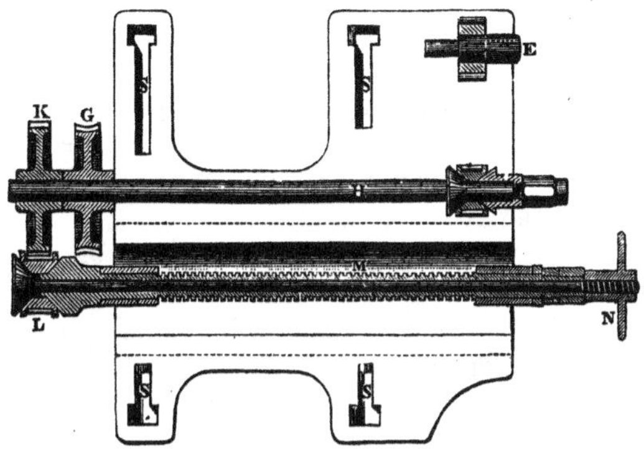

FIG. 148.

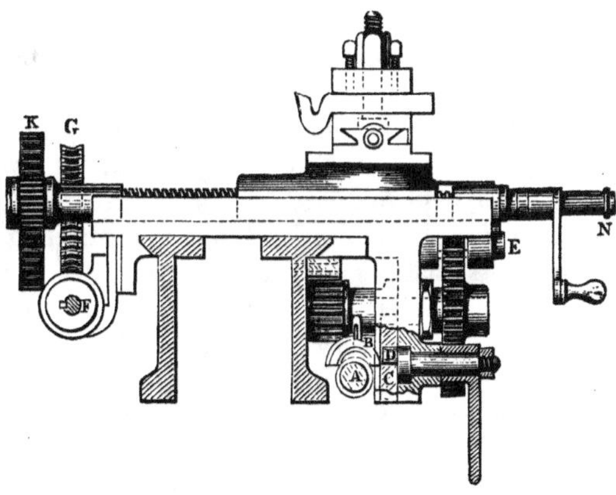

FIG. 149.

inasmuch as its movements, with the exception of that derived from the leading-screw, are only producible by hand. In Fig. 149, then, it will be observed that the half nut (B) is attached to a vertical slide (C), which is there shown partly in section, so as to exhibit the eccentric pin (D) by which it is worked. When therefore the lever handle is in the position which it there occupies, the half nut is raised clear of the leading screw (A); when on the other hand it has been moved to that in which it appears in the perspective view (Fig. 147), where the lower part of the slide may be noticed projecting below the vertical portion of the 'saddle,' the screw and the nut are brought into contact.

But except for cutting screws, it is preferable—as we have before explained—to give this longitudinal, or 'traversing' motion, by means of the rack, instead of the leading screw. In order to do this by hand, it is only necessary to turn the winch-handle on the extreme right of the rest (in Fig. 147), which fits into the socket of the spindle (E, in Figs. 148 and 149), and thus drives a pinion gearing into the spur-wheel, which appears in the first and last of these figures. At the back of this wheel is another pinion gearing into the rack. To make this traversing motion automatic, all, therefore, that is necessary is to drive the spur-wheel from the shaft at the back of the bed; which is easily effected through the medium of the worm-wheel (G), and the shaft (H), which carries a pinion (J) gearing with the spur-wheel. Ready means must, however, be provided for disconnecting the pinion (J) from its shaft, and thereby arresting the motion of the slide-rest; and the method of doing this—which obtained for Messrs. Shepherd and Hill a well-earned prize medal at the Great Exhibition of 1851—is one of great simplicity and beauty. The pinion, which is bored so as to fit loosely upon the shaft (H), is deeply countersunk on both sides, and is placed loose against a conical shoulder fixed upon the shaft. The conical face of this shoulder fits into the countersunk pinion on one side, and the face of a loose conical collar fits into it on the other. On drawing these two cones together, therefore, by means of a nut working on the screwed end of the shaft (H), the pinion becomes firmly gripped, and is compelled to revolve with it; whilst on slackening the nut (which appears in front of the rest in Fig. 147), it is instantly released.

A precisely similar arrangement is adopted for throwing in or out of gear the pinion (L), by which this rest can be made self-acting for 'surfacing' This pinion receives its action from the spur-wheel (K), which is keyed upon the shaft (H) just behind the worm-wheel (G). When clipped between its cones, it drives the cross-slide screw (M) either backwards or forwards according to the position of the rocking-frame upon the end standard of the lathe, as, it may be remembered, was pointed out when we were describing, with the assistance of Fig. 144, the means by which the back shaft itself is driven. But, in the present case, the pinion is at the back of the lathe; and since it would be highly inconvenient to have the nut there also, the cross-slide screw is made hollow, and a rod, attached to the hinder cone, is passed through it to a T-handle (N) in front. The nut at the end of the shaft (H) was till lately provided with a similar handle, but a hexagonal nut, which can be readily turned with a spanner, has been found to be more convenient.

In turning up a long shaft or any similar piece of work of which the diameter is insufficient to give the necessary rigidity, a back-stay is frequently attached to the saddle, so that it accompanies the slide-rest in its traverse, and is always favourably placed for giving the requisite support. The T grooves marked S are for the purpose of enabling this to be bolted down to the saddle.

In the Duplex lathe by Sir Joseph Whitworth, of which we have given an engraving on a former page (Fig. 143), the principle of making the whole of the upper part of the slide-rest in duplicate is adopted. The two ends of the cross-slide then have their threads running in opposite directions—one being right-handed and the other left-handed—so that the two tool-holders advance towards the axis of the work or retreat from it simultaneously. It is of course necessary to reverse the position of the tool or tools at the back of the lathe; in these, as in other slide-rests, two being frequently fixed in each tool-holder, of which one gives a roughing and the other a finishing cut. The great saving of time and labour which has resulted from the introduction of the 'Duplex' system, has led to its almost exclusive adoption at

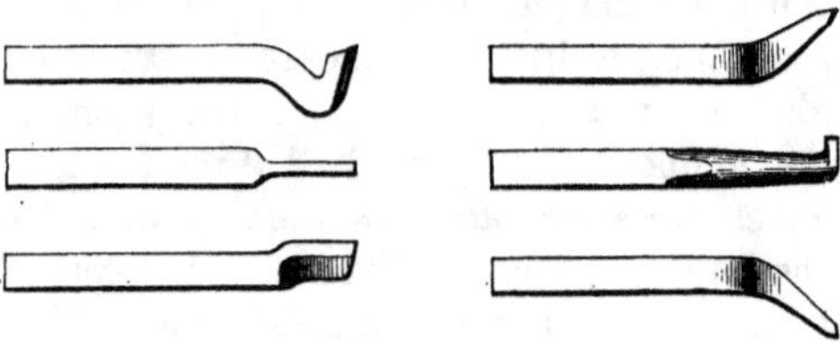

FIG. 150.—Slide-rest Tools for Power Lathes.

the works of the inventor. The fact of the strains upon the work being balanced when its two opposite sides are attacked equally, is undoubtedly much in its favour.

A few slide-rest tools of the ordinary types which are used for turning in power-lathes are represented in Fig. 150, but they are of course liable to much variation to enable them to meet the requirements of special cases. Commencing with the uppermost or the left-hand side of the figure, they are called respectively a hook-tool, a parting-tool, a knife-tool, a right-hand side-tool, an internal tool, and a left-hand side-tool; and they sufficiently explain themselves. Their shanks are one inch square in each case.

In the application of slide-rest tools with loose cutters to power-lathes, some progress has lately been made by Mr. Ford Smith, of Manchester. The form of tool which he recommends instead of the hook-tool above shown—with which the bulk of the work is ordinarily done—is given in Fig. 151. Its chief advantage over those of which engravings have already been given (in Fig. 131), is, that the loose cutter, as seen in plan, is inclined to the centre line of the shank at an angle of 45°—a position which is more favourable to it for ordinary traversing or surfacing, whilst at the same time it enables the tool to be used in lieu of a side-tool when required. Mr. Smith has found that in general—that is to say, when the depth of the cut does not exceed half the diameter of the cutter—steel of circular section is to be preferred for it, the surface of the work being left smoother than when an angular tool is employed; but for very deep cuts he adopts an oval section. He also makes use of templates, both to ensure the cutting edge being placed on a level with the axis of the mandril, and to maintain the

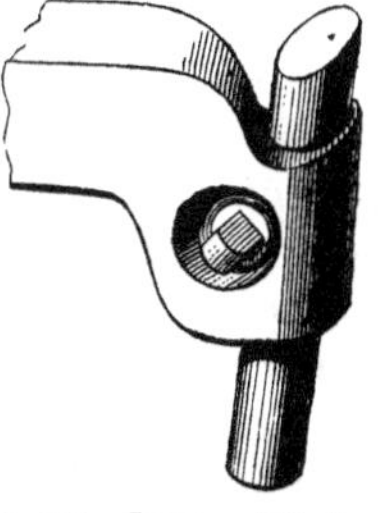

FIG. 151.—Improved Cutter-holder.

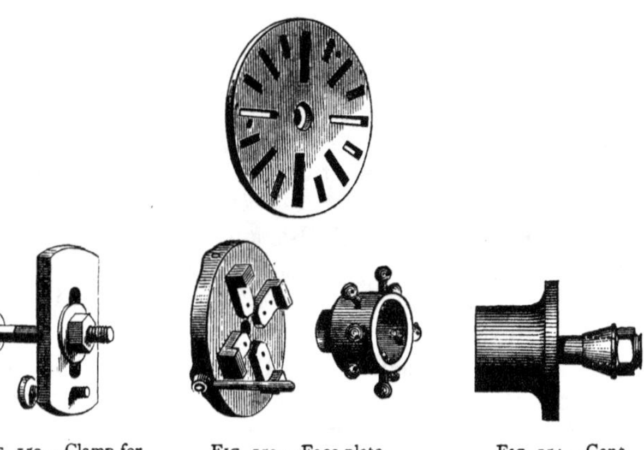

FIG. 152.—Clamp for Face-plates. FIG. 153.—Face-plate. Four Jaw Chuck. Bell Chuck. FIG. 154.—Cone Chuck.

proper angles in grinding the cutters; these angles being for wrought iron 50°, and for cast iron and brass 60°. The holder is made so as to give the face of the cutter an inclination of 3° from the perpendicular.

Of the *Chucks* by which the various articles which have to be treated in power-lathes are attached to them, one or two examples may have been noticed in our engravings of the lathes themselves. In the Break-lathe (Fig. 145), a ponderous *Face-plate* serves for this purpose, and it may be supplemented when necessary by a 'Driver' or by an *Angle chuck* (which is an L-shaped casting, planed and slotted); but this, unlike the generality of chucks, is attached permanently, so that it may be considered to form part of the lathe itself. Moveable face-plates (one of which is represented above in Fig. 153) are of much smaller dimensions. For heavy work they are used more than any other kind of chuck, since a large proportion of the articles which have to be attached to the mandril admits of either being bolted directly down to the face-plate or of being held upon it by means of clamps—one form of which is shown to a large scale in Fig. 152.

The *Four-jaw chuck*, of which a representation is also given (Fig. 153), may likewise be recognised as having appeared in a previous engraving (Fig. 146). Its use is sufficiently obvious—the jaws being separately tightened up by means of a box spanner till the work is held concentrically between them.

Similarly the *Bell-chuck*, which completes the assortment in Fig. 153, admits of any piece of work of smaller diameter being held firmly and true between the ends of the screws with which its sides are studded.

Another chuck which is useful for holding articles of small width, although they may be of large diameter, provided only they have a central hole of appropriate size, is the *Cone Chuck* (Fig. 154). The central pin, which is firmly fixed in the chuck, is passed through the hole in the work. The loose conical ring is then slipped on, and the nut screwed down upon it till the work is secured. In this way the outer edge of the hole through the work is compelled to be concentric with the chuck.

A large number of articles, however, in power-lathes, as in foot-lathes, cannot be held by the mandril only, but require also the support of the back centre. 'Drivers and carriers' by which work of this kind may be made to revolve in the lathe have already been mentioned, and engravings of them have been given (in Figs. 114 and 115); and these —altered only in respect of size—are used also in power-lathes. But for heavy work and for turning long shafts an improved driver, invented long ago by Mr. Clement, of

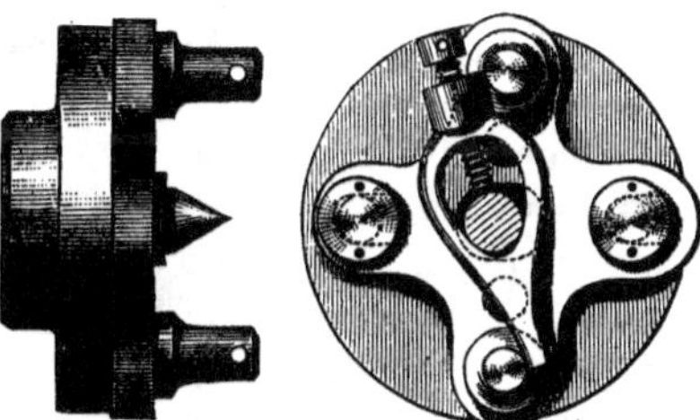

FIG. 155.—Clement's Driver (⅛).

Lambeth, which is known as *Clement's Driver* (Fig. 155), has been widely adopted. It differs from one form of the ordinary drivers in having two driving pins instead of one, and in having them fixed in an outer plate instead of in that which is attached to the mandril. This outer plate is capable of sliding laterally for a short distance, thus accommodating itself to any inequality in the width of the opposite ends of the carrier; by which means each of the pins is made to transmit an equal share of the driving power.

CHAPTER VII.

ON DRILLING AND BORING MACHINERY.

IN a previous chapter (Chapter III.) we noticed briefly some of the portable forms of apparatus with which holes of moderate diameter (up to about an inch) can be made in wrought iron and similar materials, by giving a slow revolving motion to a drill with two strong cutting edges. This, as we saw, is generally done either by a simple crank worked by hand, as in the smith's brace, or by converting the reciprocating motion of a lever into one of intermittent revolution, as in the numerous kinds of ratchet brace. In either case the process is of necessity a slow one, as the power available for overcoming the great resistance which the drill encounters is limited to that which can be applied by the arm of the workman. Consequently, although these appliances are most valuable for drilling the comparatively small number of holes which must be made during the erection of any iron structure, or for those which are required in such parts of a piece of work as are difficult of access, they are but seldom employed for workshop purposes, for which *drilling machines* of greater or less power are much superior in point of speed and accuracy of performance.

The chief qualifications essential to a drilling machine which is to be used for miscellaneous work are as follows: First, it must be capable of being readily connected with the shafting by which the driving power is transmitted to the various parts of a workshop, and in such a manner that the speed of the drill can be varied; secondly, it must be provided with an efficient and variable 'feed motion;' and, thirdly, it must have a perfectly firm 'table' for the reception of small articles, which must offer as little obstruction as possible to large ones. Stability and strength of framing are of course most important qualities for all machine tools, though they are not invariably to be found in the frames of drilling machines.

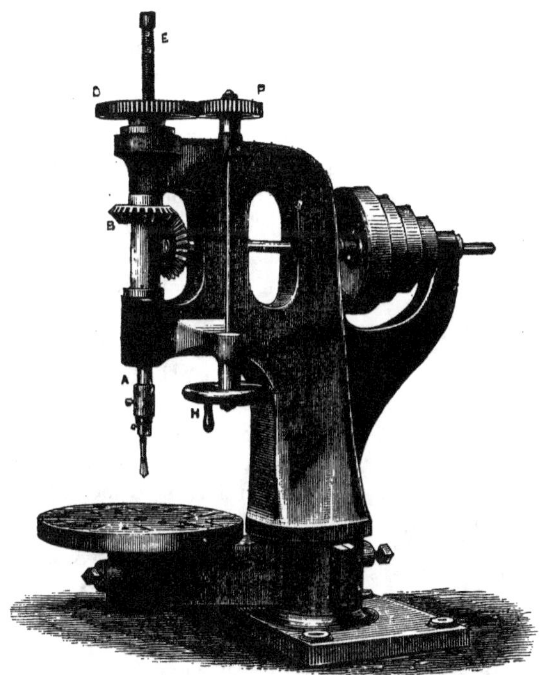

FIG. 156.—Single-Geared Drilling Machine.

Fig. 156 shows a *Single-Geared Drilling Machine*, by Messrs. Shepherd, Hill, & Co., of which the upper part is a good type of these machines in their simplest form, though the lower has some peculiarities, which we will notice directly. In all ordinary drilling machines the *spindle* which holds the drill is placed vertically, so that a short horizontal shaft, connected with it by bevil-wheels, is necessary to enable it to be driven by a belt from any main line of shafting, or from a countershaft, these being invariably horizontal. The conical *speed pulley*, which, by being driven—as in the case of a power-lathe—from a similar pulley with the cone reversed, enables considerable variation of speed to be obtained, is sometimes placed within the framing, so as to bring it much closer to the bevil-wheels, part of the framing being then above it ; but this arrangement, though somewhat more compact in appearance, has the disadvantage of almost always requiring the machine to be erected vertically under and parallel to the driving-shaft, to enable the belt by which it is driven to run clear of the framing.

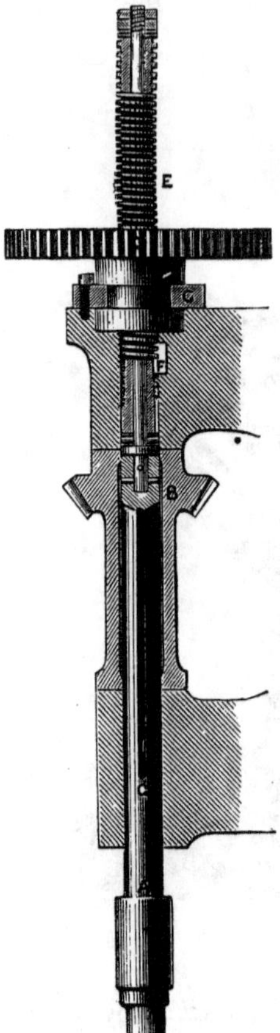

FIG. 157.
Section of Drill-Spindle.

This framing, and also that of the 'radial drilling machine,' which will be found a few pages farther on, should be noticed, being good examples of the excellent system of casting them hollow and in one piece, which was introduced some twenty years ago. To turn out these complicated castings without defects, and of uniform thickness, requires no small amount of skill in the moulder's difficult art.

This machine is provided with a 'screw-feed,' worked by hand, of which the general arrangement appears with great distinctness in the above engraving ; the effect of turning the hand-wheel (H) being obviously to raise or lower the drill-spindle. But the construction of the several parts—which is not quite so simple as might at first be supposed—will be rendered more intelligible by the section (Fig. 157), in which the letters refer to the same parts as in the preceding figure. The spindle (A), the lower end of which receives the drill, slides freely through the elongated boss, or 'pipe,' of the horizontal bevil-wheel (B). Inside this boss a feather, fitting loosely into a groove (G) upon the spindle, compels it to revolve with the bevil-wheel without interfering with its independent vertical motion. But this vertical or 'feed' motion is controlled by the long screw (E), which is so connected with the upper end of the drill-spindle (A) as to be able to raise or lower it, either when at rest or when revolving, without itself sharing the revolution; its tendency to do so being checked by a longitudinal groove and a feather (F) fixed within the framing. In order to raise the point of a drill fixed in the spindle (A), we have therefore merely to draw the screw (E) upwards by means of any convenient form of nut; the most convenient in this case being a spur-wheel (D) with a female screw cut through its boss, which can then be worked by a pinion (P) and hand-wheel (H), within easy reach of the workman. But besides raising and lowering the drill, it is necessary to be able to put considerable downward or 'feed' pressure upon it. This can be done by merely continuing to turn the hand-wheel (to the right, the feeding screw being 'left-handed'), after the drill has been brought into contact with the work, the spur-wheel being prevented from rising by a split collar (C), which fits into a horizontal groove on the outside of its boss, and is bolted down to the framing.

The advantages which, as we have seen, result from making power-lathes 'self-acting' are almost equally great when the same principle is applied to drilling machines. With the screw-feeding arrangement above described, it is simply necessary to impart to the hand-wheel a continuous slow motion, derived from some neighbouring moving part, to cause the drill to cut with perfect uniformity, instead of at one moment making little or no progress, and at the next having its strength unduly taxed through the inconstant attention of a workman. In the *wall-drill* (Fig. 158) the mode in which this is usually effected can be seen. Running loose upon the vertical shaft of the hand-wheel (H) is a

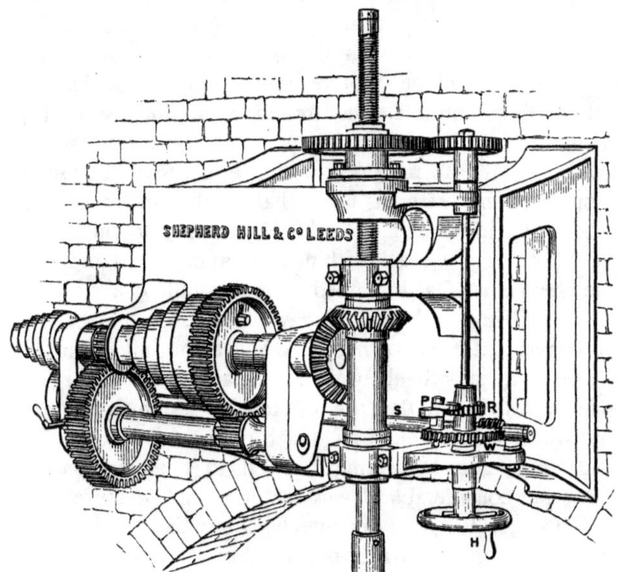

FIG. 158.—Double-Geared Wall Drilling-Machine.

worm-wheel (W), which is constantly being driven by an endless screw fixed upon a light horizontal shaft (S); this shaft being connected with the horizontal driving-shaft by a short belt passing over a pair of small speed-pulleys. Keyed upon the vertical shaft is a ratchet-wheel (R), which can at any moment be made to revolve with the worm-wheel (W) by means of a pawl (P) carried by the latter on its upper surface. On releasing the pawl, therefore, the drill can be raised or lowered by turning the hand-wheel (H); and on causing it to engage with the ratchet-wheel, the self-acting feed is brought into play, its amount being capable of being regulated by shifting the strap on the speed-pulleys.

But several other kinds of self-acting feed have been devised. One, invented by Sir J. Whitworth, is an expedient of considerable elegance, which Fig. 159 will serve to explain. Upon the upper part of the drill-spindle—which in this case is in one piece throughout its length—a screw-thread of moderate coarseness is cut, the somewhat complicated joint between the screw and the spindle shown in Fig. 157 being thus rendered unnecessary. If a fixed nut were attached to the upper part of the framing, this screw would provide the drill with a self-acting feed; but it would have to be always exactly equal to the pitch of the screw, and could therefore never be altered. So, instead of letting it run through a fixed nut, it is made to pass between two worm-wheels, which, whenever the vertical motion of the spindle is arrested, themselves revolve in opposite directions. (One only of these worm-wheels can be seen in the engraving.) Their revolution, however, can be either wholly or partially checked by turning a handle, which works a right- and left-handed screw, and so draws together two clips (CC). These clips embrace a pair of friction-wheels (FF), keyed upon the short shafts which carry the worm-wheels, one of the shafts also having upon it a hand-wheel, which is not shown in the figure. When the friction-wheels are so tightly clipped that they cannot revolve at all, they have the same effect as a fixed nut, and the feed is equal to

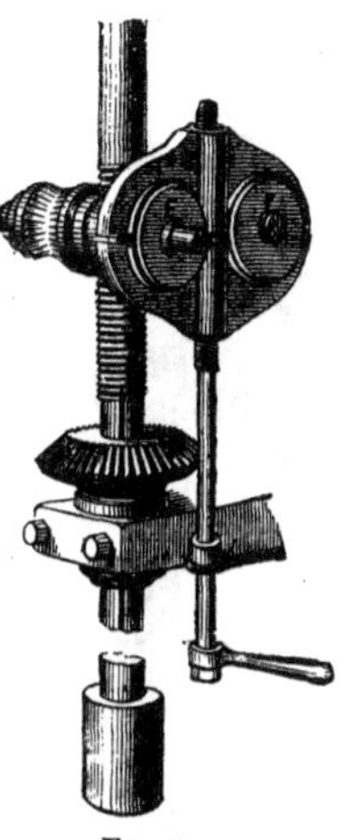

FIG. 159.
Whitworth's Drill-Feed.

the pitch of the screw; but for each diminution in the tightness of the clips the pressure upon the drill is reduced, and an increased motion is given to the worm- and friction-wheels. By loosening the clips and turning the hand-wheel, the drill-spindle (which is provided with a counterpoise to keep it from falling by its own weight) can be quickly raised or lowered.

An ingenious but an invariable continuous feed motion, applied by Mr. Bodmer, of Manchester, is described by Professor Goodeve in the volume of this series previously mentioned,[1] and is represented in Fig. 160. The drill-spindle—like the previous one—has a long thread cut upon the upper part of it, and screws into the boss of a spur-wheel, which thus forms a nut, which can itself be made to revolve. If both the spindle and the nut are driven at the same rate, the drill of course receives no feed motion. But by driving them in the direction indicated by the arrow, and giving them slightly different speeds, the spindle is slowly and continuously unscrewed; the amount of its descent (or feed) during each revolution depending upon the excess of its speed over that of the nut. Thus if, during each turn of the spindle, the nut passes through only three-quarters of a revolution, the feed will be equal to a quarter of the pitch of the screw.

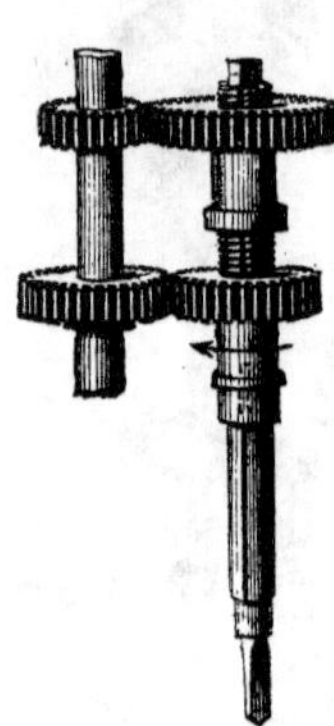

FIG. 160.
Bodmer's Drill-Feed.

Another kind of continuous feed, which is more extensively used for large drilling machines than any of the preceding ones, is on much the same principle as the rack-traversing motion in self-acting lathes. The drill-spindle is made in two parts, the lower, which carries the drill, being free to revolve independently of the upper, to which it is attached by a joint resembling that employed for connecting

[1] *Elements of Mechanism*, pp. 204, 205.

together the screw and the spindle in the screw feed machines. In the circular pillar drilling machine (Fig. 161), the gearing for this *rack-feed*, as it is called, can be clearly seen, the teeth of the rack appearing on the right-hand side of the upper part of the drill-spindle. But in order to render this kind of feed self-acting, the driving speed must be reduced to a much greater extent than is necessary in the case of the screw-feed. This is effected by introducing a second worm-wheel for communicating the motion to the pinion which gears into the rack, in addition to the lower one, which is placed above the hand-wheel. The latter receives its motion from the smaller pair of speed-pulleys, by which the rate of feed can be varied; this portion of the arrangement exactly resembling that already shown in Fig. 158.

In addition to the feeding apparatus of drilling machines, much ingenuity has been expended in devising convenient forms of support for the work upon which they have to operate. One method—and the simplest one—is to extend the foot of the frame of the machine to a considerable distance, so as to make it form a flat *base-plate.* T-grooves for the reception of the heads of holding-down bolts being cast in this plate, and its surface being planed, it forms a very efficient support for articles of moderate size of which the height does not exceed the distance between the point of the drill and the surface of the base-plate; for when a fixed base-plate is immediately under the spindle in any drilling machine, it of course sets a rigid limit to the height of the work which can be admitted. An example of a base-plate will be found in the engraving of the radial drilling machine, Fig. 163. Being of large size, it is cast in a separate piece from the framing; and, owing to the horizontal movement of the drill-spindle, the height of the work is not in this case limited by it.

Nor is this limit in point of height the only disadvantage of a fixed base-plate. Being kept as low as possible for the reception of large articles, it becomes a very inconvenient supporter of small ones. For these a *table*, adjustable as to height, is much better adapted. In its simplest form, it consists of a rectangular plate, planed and provided with T-grooves, which can be raised or lowered by a rack and pinion—the latter driven by a worm and worm-wheel, or in any other convenient way. But in this case the work must be perfectly correctly placed as to its horizontal position before it is bolted down to the table; and in drilling a series of holes, it must be set free; and be again adjusted for each one of the series. To obviate this, Sir J. Whitworth introduced his *compound table*—giving to the ordinary table a horizontal motion, on the principle of the slide-rest, in one direction, and, in addition, hanging it at a point near the vertical portion of the framing, so as to enable it to be swung horizontally in the other. The hinge also made it possible to swing the table entirely out of the way when the work was of sufficient height to be supported on a base-plate.

Various modifications of this are now in use, under the name of *swing tables*, the advantages of the arrangement for small work being universally recognised. For machines of comparatively small size, that shown in Fig. 156 leaves, perhaps, nothing to be desired. The table, which is circular, and is free to revolve round its axis, is supported by an arm, which can itself revolve round the vertical pillar of the machine. By combining these two motions, every part of the table can be easily brought under the drill, so that any piece of work of moderate size fixed upon it (for which purpose it has a series of slots, like the face-plate of a lathe) need not be unfastened till all the vertical holes which have to be drilled in it have been completed. In this case no vertical adjustment is provided, the table being placed at a convenient height for holding moderately small articles, and both table and arm being swung completely out of the way of any large ones. There being no base-plate, there is no particular limit to the height of the work which can be operated upon, a pit being occasionally sunk at the foot of the machine for the reception of long pipes, &c.

In larger machines—for the tables of which a vertical adjustment is important—the following arrangement is sometimes used, and it is an expedient of considerable ingenuity:—The pillar by which the table is carried is converted into a 'circular rack,' by turning upon it a series of grooves. Into these grooves gears a pinion, just as it would gear into an ordinary rack—with this difference only, that whereas its teeth would engage with an ordinary rack only when the arm of the table in which it is fixed was on one side of the pillar, with the circular rack it can do so equally well on all sides. This enables the table to be raised or lowered, whilst at the same time it is slewed round to any required extent—providing it, in fact, with an efficient vertical adjustment, without in any way interfering with the horizontal ones just described.

Another way of effecting this, which admits of a plain pillar being substituted for the circular rack above mentioned, is shown in the *circular pillar drilling machine*, Fig. 161. The arm on which the table is supported embraces not only the pillar, but also an ordinary straight rack, which it carries round with it to any side of the pillar to which it may be swung. The worm and worm-wheel, which enable the height of the table to be adjusted by hand, may be clearly seen in the engraving.

With reference to the methods of fixing work upon the tables or base-plates of drilling machines, the reader who is familiar with the use of the face-plate of the turning-lathe will at once recognise the object of the slots with which the circular tables in Figs. 156 and 161 are provided. Any article which has vertical bolt-holes, horizontal flanges, or which is itself of no great height, can be secured by passing either plain or hook-bolts through the slots in the table; or clamps resembling those for the lathe face-plate (shown in Fig. 152) can be used when necessary. And upon rectangular tables or base-plates pieces of work can be secured on a similar manner, by means of their T grooves, into any part of which holding-down bolts can be inserted. But in most cases the weight of the work is sufficient to steady it, and to prevent its revolving with the drill; and if it manifests

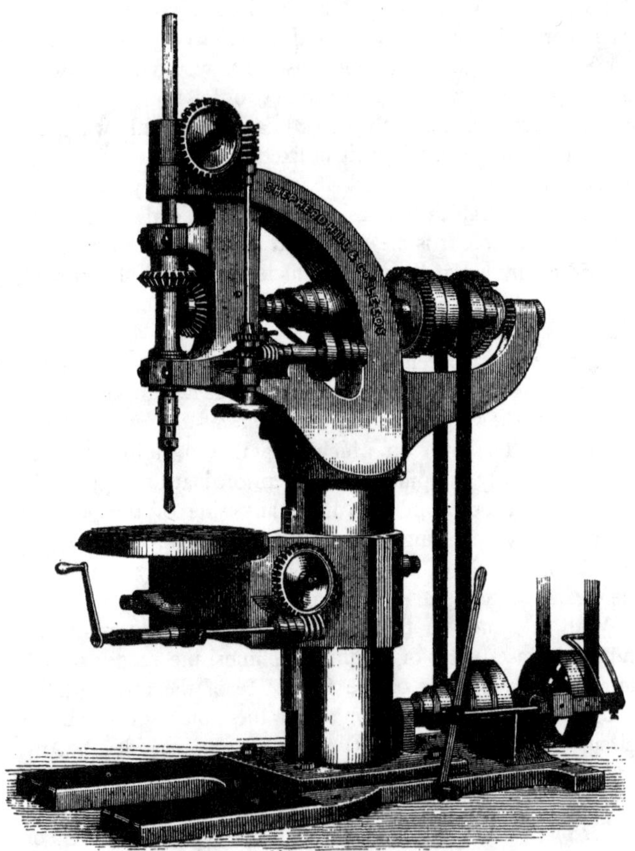

FIG. 161.—Double-geared Pillar Drilling Machine.

any disposition to move, the workman is generally able to steady it with his hand, so that it is unnecessary to secure it with bolts or otherwise. Large plates, &c., which cannot rest wholly upon the table, require of course extraneous support; but the shortness of the time occupied in drilling any moderate-sized hole generally makes it preferable to hold them by hand rather than to have recourse to any mechanical means of fixing them.

For steadying small pieces, especially such as have not a firm basis of their own—as, for instance, in drilling a hole transversely to the axis of any cylindrical work of small diameter—a *vice-chuck*, like Fig. 162, is a convenient addition to a drilling machine. The engraving sufficiently explains it—the slots in the base being for the purpose of bolting it

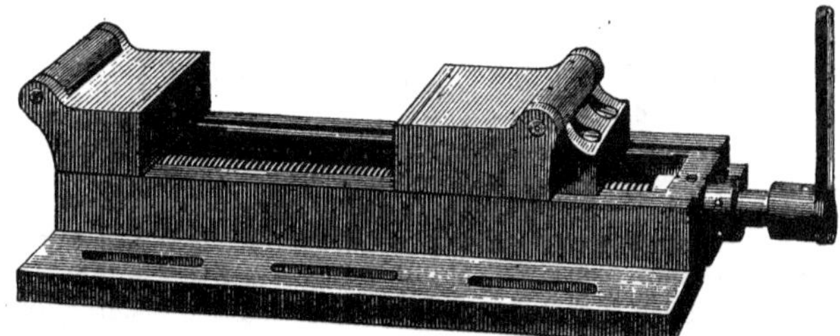

FIG. 162.—Vice-Chuck.

down to the table, and the lever (which is moveable) for opening or closing up the jaws by means of the internal screw. The centres upon which the jaws are hinged should, however, be noticed, as their position, well above the point at which each jaw transmits its pressure to the work, conduces much to the steadiness of any article within its grasp.

For holding articles whose form precludes their being bolted down upon a horizontal surface, but which can be attached to a vertical one, an *angle-chuck* is generally supplied with vertical drilling machines. It consists, like the lathe-chuck of the same name, of a short bar of L-, or angle-iron, planed on its outer faces. Both its horizontal and its vertical portions contain slots for the reception of bolts,—in the one case for fixing the chuck to the table or base-plate of the machine, and, in the other, for attaching the work to the chuck.

When a drilling machine is required for operating upon large pieces of work only, a wall-drill—of one example of which an engraving has already been given (Fig. 158), and in which both base-plate and table are dispensed with—is perhaps the most convenient form of all. Its framing is then so made that it can be fixed to the wall of the workshop, and the whole of it is kept above the drill, so as to be quite clear of any work placed underneath it. Moreover, by turning an arch in the wall, and fixing the machine at its crown, the clear space below the framing can be still farther extended, so that a boiler or any other large object can be operated upon with perfect facility.

'Double gearing' has been so fully described in connection with the lathe, that it is unnecessary to do more than to refer to it here. Both its object—that of obtaining greatly increased power by a corresponding sacrifice of speed—and the gearing by which this is effected, are in general perfectly similar, whether the work is to be performed in a power-lathe or in a drilling machine. The latter, when intended for drilling holes of more than about two inches in diameter, should always be provided with it.

One other variety of vertical drilling machine should not be passed over, inasmuch as it affords facilities for drilling vertical holes in large or ponderous articles by adopting the principle of adjusting the drill over the work instead of arranging the work under the drill. In this respect the *radial drilling machine* (Fig. 163) differs materially from the machines for drilling which we have thus far been considering. The way in which this power of adjustment is obtained—as will be understood from the engraving—consists in supporting the drill-spindle, together with the gearing for driving and feeding it, upon a slide or carriage which can travel horizontally along a radial arm, the arm itself being moveable about a vertical hinge near its point of junction with the fixed pillar of the framing. By combining the travel of the slide with the circular motion of the arm, the drill can be brought not only over any part of the base-plate of the machine, but even to a considerable distance beyond its limits on either side. The height of the arm can also be

FIG. 163.—Radial Drilling Machine.

adjusted, it being for that purpose hinged to a long vertical slide, instead of directly to the framing.

But these advantages are not obtained without considerable sacrifice of simplicity in the working parts, as must be evident when we consider that the driving power has to be transmitted first to the radial arm, whose height and direction are variable, and from it to the drill-spindle, of which the position and height are also variable. Means must be provided too for raising or lowering the arm, and for regulating the traverse of its slide, without interfering with the self-acting feed-motion of the drill-spindle. The manner in which these requirements are satisfied will probably be self-evident on studying the figure, in which the reader who has accompanied us thus far in the subject of drilling machinery will not find much difficulty in tracing out for himself the course of the power from the driving-pulleys to the drill, nor in understanding what may be called the subsidiary parts—for feeding, adjusting, &c.

One fact should be borne in mind, for it greatly simplifies this task, not only in machines of comparatively few parts, like the present one, but in most others, however complicated they may at first sight appear. It is this : that in all properly designed machinery the strength of the working parts is proportioned to the strains they have to bear; so that by observing their strength a good idea may be obtained as to the amount of power they are intended to transmit. Toothed wheels, shafts, &c., will thus be found to a great extent to tell their own story, the thickness of the teeth of the former and the breadth of their faces, and the diameter and length of the bearings of the latter, being frequently most valuable guides in assigning to each their respective duties. Examining, for instance, in this way the various working parts of Fig. 163—keeping in mind of course the main object of the machine, which is in this case to convey the power from the driving-pulleys to the drill-spindle without any great alteration of speed (except through the agency of the double gearing)—a single glance assures us that the horizontal shaft running along the radial arm, and also the vertical shaft between the trunnions on which the arm is hinged, are for the purpose of transmitting this power, since they, and also the drill-spindle, are approximately of the same diameter. An inspection of the gearing by which they are connected confirms this impression.

The vertical shaft is driven from the double-geared pulleys by a pair of strong bevil-wheels—which are clearly seen near the centre of the hinge of the radial arm—the rise and fall of the arm being provided for by grooving the vertical shaft and fixing a feather in the boss of its bevil-wheel, this wheel being supported by a bracket projecting from the fixed part of the framing. Similar bevil-wheels connect the vertical with the horizontal shaft (one of these is partly seen in the figure just below the upper trunnion of the radial arm), which also is grooved throughout its length on account of the traverse of the slide, a pinion in this case at the back of the slide being carried with it to any part of the arm to which it may be moved, and a feather in the boss of the pinion remaining constantly in the groove of the shaft, so as to receive its motion. This pinion cannot be seen in the engraving, but the spur-wheel in front of the slide which gears into it shows us at once by the width of its face (which is almost equal to that of the spur-wheels of the double gearing) that it transmits the driving power to the drill, and has nothing to do with its adjustment or feed. One other pair of bevil-wheels gives revolution to the drill-spindle—*which* of the two pair with which it is furnished the reader will have no hesitation in determining for himself on the principles above recommended. The others convey the minute proportion of the driving power which is required to make the feed self-acting; a pair of small conical pulleys at the back of the radial arm giving the requisite control over its speed. These, however, and the 'rack-feed' generally, have already been noticed; so we will here only call attention to the smallness of the diameter of the vertical shaft, which performs the very light duty of connecting the lower worm-wheel with the upper worm. Still one other pair of bevil-wheels remains, namely, those in front of the standard of the framing. They are for the purpose of raising or lowering the radial arm, together with the slide, gearing, &c., which it carries; the combined weight of which renders it necessary to give considerable strength to the teeth of the wheels by which this is effected. So that in this case there may at first sight be some doubt as to their use, although a small amount of consideration (with which the application of what has been said above must always be accompanied) soon points out the necessity for their somewhat exceptional strength. In this adjustment a hand-wheel, worm, and worm-wheel are sometimes substituted for the bevil-wheels and pulley shown in the engraving, the circular motion being in each case converted into a vertical one by means of a pinion gearing into a rack which is fixed to the slide upon which the radial arm is hinged.

The bases of radial drilling machines are occasionally made of sufficient height to enable their vertical sides to be utilised in holding articles which can be more conveniently affixed to a vertical than to a horizontal surface. When this is the case the base assumes the form of a large rectangular box, with horizontal T-grooves cast upon its sides and ends.

The following particulars of the vertical drilling machines figured above will serve as a guide in stating the capabilities of others of the same kind.

The single-geared machine (Fig. 156) is capable of drilling holes up to 1½ inch in diameter, to a depth of 10 inches, the distance of the centre of the hole from the edge of the work being limited to one foot.

The wall-drill, of which an outline is given in Fig. 158, has, of course, no limit as to the position of the hole which it can drill, and being double-geared, its diameter can be as much as 6 inches, the travel of the drill-spindle, which determines its depth, being 14 inches.

The powerful double-geared 'circular pillar' machine (Fig. 161), can, like the preceding one, drill holes up to 6 inches in diameter, their greatest distance from the edge of the work being 15 inches. It can admit articles 3 feet 6 inches high.

The radial drilling machine (Fig. 163) is also adapted for

drilling holes up to 6 inches in diameter, the travel of the spindle being 13 inches. Its arm has 18 inches of vertical adjustment, and can be swung round upon its axis through something more than a semicircle. The travel of the slide upon the arm enables the drill to be placed at any distance between 2 feet 6 inches and 6 feet from the centre of its trunnions.

The ordinary form of *drill* which is used with these ma-

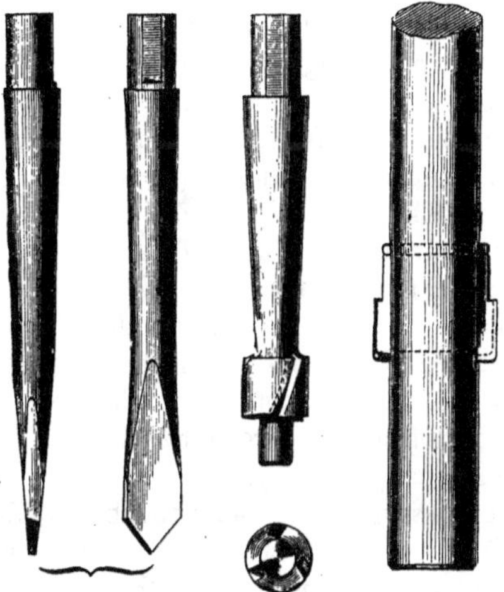

FIG. 164.—Common Drill and Pin-Drills.

chines, together with two kinds of pin-drill, is represented in Fig. 164. One much resembling the first of these has already been given in connection with hand-drilling apparatus, but greater attention is naturally paid to the proportions of these tools when they are of large than when only of small size, and of these the present is a good example. The parallel portion above the cutting edge should be noticed, since, as already pointed out, it conduces much to the regularity and smoothness of the hole produced. The *Pin-*

Drills speak for themselves; the first, with three cutting edges, being used for such purposes as enlarging the ends of screw-holes, so as to admit the heads of 'cheese-headed' screws, and being sometimes called a 'recessing bitt.' The second is a tool of a good and much used type, consisting of a loose steel cutter, held by a key in a bar of wrought iron or steel of appropriate length. When of any considerable size this is known as a 'boring bar,' and it is then frequently used in some form of (generally horizontal) boring machine by which both its extremities can be supported, instead of in such drilling machines as we have been considering. Into the construction of these we cannot enter, but a brief description of a boring machine of much larger size, in which the feed motion is given to the cutter instead of to the bar itself, will be found a few pages farther on.

But before quitting the subject of simple vertical drilling we must mention the so-called *Multiple Drilling Machines.* Whatever may have been the purpose for which they were originally devised, their first introduction on a large scale was due to Messrs. Cochrane of Dudley, who employed them in the manufacture of wrought-iron girders. For such purposes—in which it is necessary to drill very large numbers of rivet-holes with the utmost possible rapidity, and to place the drills with such accuracy as to ensure perfect correspondence between the holes—they offer very great advantages. The end standards, with a portion of the horizontal girder by which the drill-spindles are carried in a machine of this kind made by Messrs. Collier of Manchester, by which upwards of forty holes can be drilled simultaneously, is represented in Fig. 165. A section through the girder is also given to an enlarged scale. The standards are placed at a sufficient distance apart to admit between them the longest plate which has to be operated upon, so that the length of the machine varies according to the purpose for which it is intended. In any case a planed table for the support of the plate extends under the whole range of the drills, hydraulic cylinders or other mechanical means being provided for slowly raising this table, and thus bringing the work gradually up to the drills, instead of complicating the spindles by giving them any vertical feed motion. One

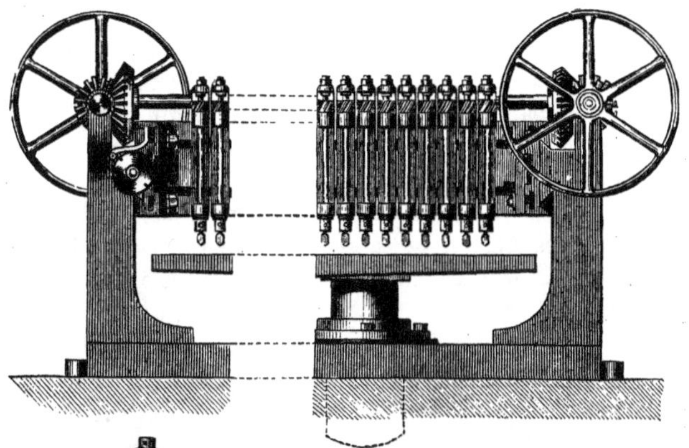

FIG. 165.—Multiple Drilling Machine. Part Elevation and Section.

hydraulic cylinder only appears in our engraving, the second, together with about three-fourths of the number of drill-spindles with which the machine is provided, having been necessarily omitted. To enable the holes in the plate to be drilled at any required points, each spindle is supported in an independent frame, which can be fixed at any part of the girder by two ordinary screw-bolts, horizontal T-grooves being cast in the girder to receive their heads. The drills when arranged are driven simultaneously from a horizontal shaft running along the top of the girder, connection being established between them by an unusual and ingenious form of helical gearing, which for such a case as this has considerable advantages over ordinary bevil-wheels. It consists in placing on each drill-spindle, and also on the driving shaft opposite to it, a short length of a many-threaded screw. These, which are precisely similar to one another, and in appearance resemble small spur-wheels with inclined teeth, are capable of working together (by sliding instead of rolling contact) when their axes are at right angles to each other. Being of much smaller size than bevil-wheels of equal strength, they allow the spindles to be placed much closer together than would be possible if they were driven in the usual way with bevil-gearing. Indeed, by driving the alternate spindles from separate shafts placed one above the other, they can be so arranged as to drill holes of which the centres are only $2\frac{1}{2}$ inches apart.

With none of the foregoing machines is it possible to drill holes of other than circular section, although for some purposes—such as in the expansion joints of iron structures, and possibly also for obtaining increased sectional area in rivetted plates generally—it is desirable to give them a more or less elongated section. With punching machines, as we shall see presently, holes of almost any required form can be produced, but for highly finished work they are inadmissible, and machines have, therefore, been devised with which elongated holes (see Fig. 162) can be drilled at a single operation, the proportions between their length and widths being variable within very considerable limits. This process is known as Slot-Drilling, and one variety of machine by which it can be performed is shown in Fig. 166. It will be seen that the drill-spindle is mounted on a carriage which is supported by a horizontal slide, much in the same manner as in the radial machine (Fig. 163). The carriage, however, instead of being adjusted once for all, as in that case, continually receives a slow backward and forward motion along the slide to an extent which depends upon the amount by which the length of the hole is to exceed its width. This, which proceeds simultaneously with the rotation of the drill, produces a hole of the desired form, which may be carried either entirely or only partially through the work. In the engraving the 'cross-head' of a steam-engine is

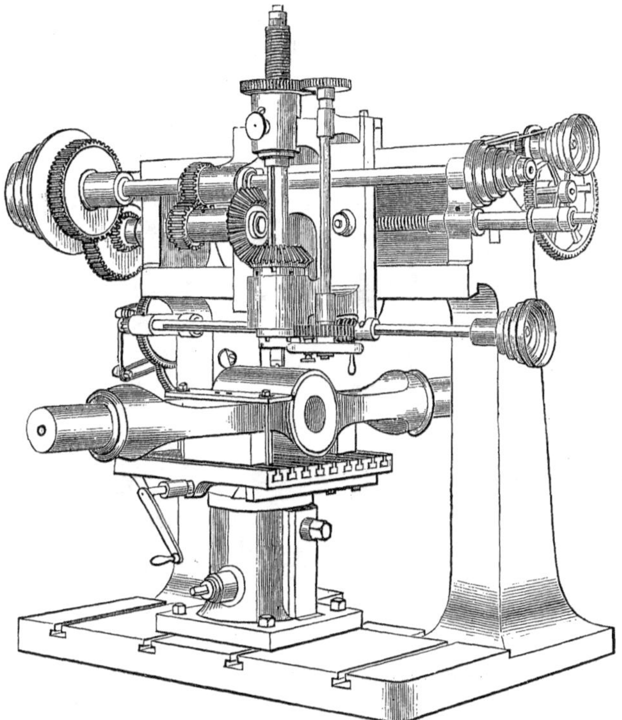

FIG. 166.—Slot-Drilling Machine.

shown in position for having a cotter-hole drilled in this manner, the table on which it is supported being capable of being adjusted as to height, or being entirely removed, so that no vertical adjustment is required for the slide. The movement is given to the carriage by connecting it at the back with a stud which projects from the face of a revolving disc, the extent of the motion being determined by the distance from the centre of the disc at which the stud is fixed. In order to overcome the irregularity with which the carriage would move at the different parts of its traverse if the disc were to revolve at a uniform rate, the pinion by which it is driven is made eccentric and the disc itself elliptical; by which arrangement the lateral movement of the

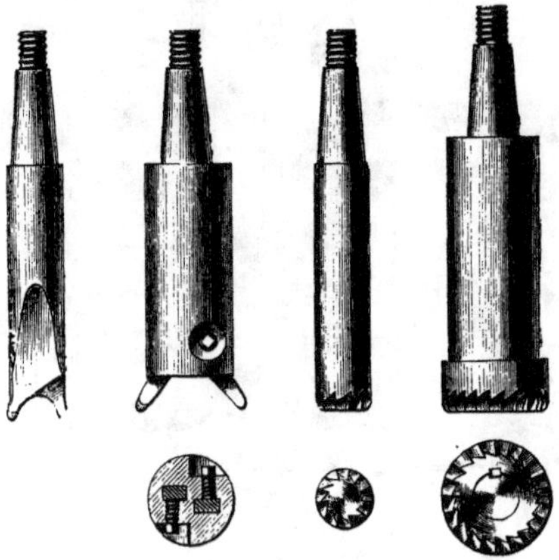

Fig. 167.—Tools for Slot-Drilling.

carriage, and therefore of the drill which it supports, is rendered everywhere nearly uniform.

For the combined horizontal and vertical cut thus obtained, the usual pointed drills cannot be employed. It has been found, indeed, that the best results are obtained by using a tool from which the central part of the cutting edge has been entirely removed, so as to give it the form of the *Forked Drill* in the accompanying figure (Fig. 167). Or a similar instrument may be made by fixing two loose cutters diagonally (also shown in the figure) in a stem somewhat less in diameter than the width of the hole to be drilled.

But these, although rapid in their action, and therefore well adapted for performing the bulk of the work, must be followed by some kind of *Finishing Drill* whenever it is required that the surface be left smooth. Two examples of Rose Drills for this purpose are given in the figure, in one of which the cutter is formed in a separate piece from the stem, and can be detached from it. But square-ended drills, from which a minute portion of the edge at the centre only is removed, may also be employed for smoothing. The stems of slotting drills, whatever their form may be, and also the spindles which carry them, must possess much greater strength than is required for ordinary drilling, on account of the side strain to which they are subjected.

The machine engraved (Fig. 166),—made by Messrs. Sharp, Stewart, & Co. of Manchester,—is capable of taking in objects of any length, provided that the height does not exceed 3 feet, and slotted holes up to 13 inches by 4½ inches can be drilled in them to a depth of 11½ inches. Various other forms are also made, some of which are provided with duplicate spindles, tables, &c., so that two slotted holes or grooves can be drilled at the same time either in one or in two separate pieces of work. For the use of these machines is by no means confined to the drilling of holes for such purposes as we have mentioned. Key-beds and many similar recesses can be cut with them in positions upon which other machine tools could not possibly be brought to bear, and their performance is such that no after-treatment with the file or other tool is required. On a similar principle—the motion, however, being given to the work instead of the tools—an entire series of slotted holes could be easily drilled at once if required.

In passing on now to *Boring*, which implies the use of some description of *Boring Bar*, we may mention that one very simple method of performing the operation consists in fixing between the centres of a lathe a plain iron bar with one or more transverse slots in it. Into each slot either a single- or a double-ended cutter is keyed, the article to be bored, which must have previously had an opening made completely through it of sufficient size to admit the bar—being attached to the saddle of the slide-rest, the movement of which along the bed of the lathe feeds the work continuously up to the tool. And for accurate work—as, for instance, in boring out a cylinder for a steam-engine—it is important for the feed to be continuous, and for the cut to be carried through from end to end without stopping, whatever may be the particular means adopted for producing it. To obtain good results the diameter of the bar must be sufficient to enable it to support the cutter rigidly and at no great distance from its cutting edge.

But for boring cylinders, &c., of large diameter there are obvious advantages in employing some form of boring bar which admits of the work being kept stationary and the feed-motion being given to the tool, especially if this is compatible with a reduction in the large size of the bar which would otherwise be required to provide the cutter with adequate support. Such an arrangement is shown in the following illustration (Fig. 168), but as the boring bar which there appears, and which constitutes the main part of the machine, may be used with almost equal efficiency in an ordinary lathe, we will bestow our attention chiefly upon it without particularly regarding the worm and worm-wheel by which in this case it is driven; which, indeed, explain themselves. The spur-wheels, however, on the left in the engraving form part of the mechanism of the bar, and must not be supposed to belong to the driving apparatus.

At the centre of the bed in the figure (Fig. 168) is a locomotive engine cylinder placed in position for being bored, and on the point of entering it may be observed a block or boss upon the boring bar. In this—which is called the 'cutter-block'—the tools are fixed, instead of in the bar itself; four or more slots being provided for their

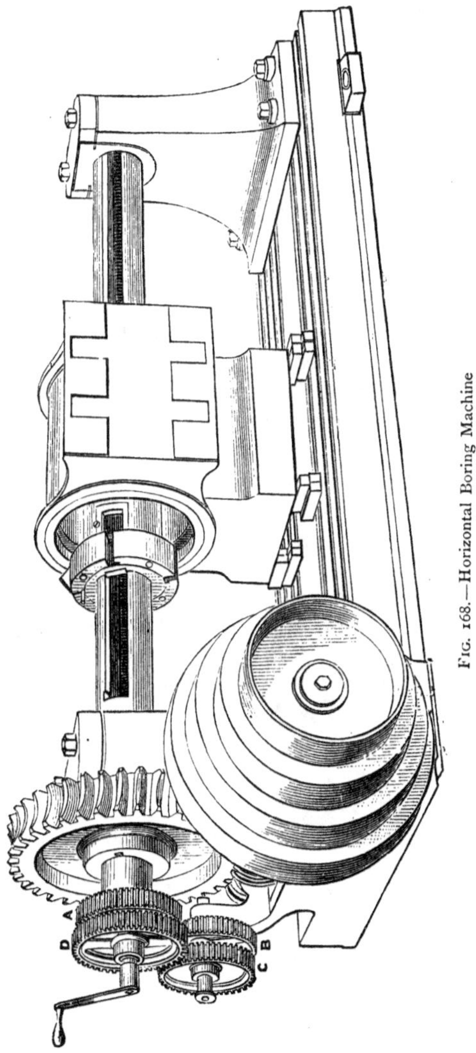

FIG. 168.—Horizontal Boring Machine

reception, together with the keys or cotters by which they are made fast. The cutter-block is capable of sliding along the whole length of the bar, so that all that is required for giving a feed-motion to the cutter is to render its movement in this direction sufficiently slow and uniform, which is always done by connecting it with a horizontal screw contained in a recess within the boring bar. But for driving this screw several different methods have been adopted. As shown in the engraving, the screw is fixed concentrically within the bar, and at one end is carried beyond it and through the centre of a spur-wheel (A) which is keyed upon the bar. Upon the end of the screw a second spur-wheel (D) is fixed, its diameter being greater than that of the first by an amount which depends upon the speed which is to be given to the screw. Into these gear two wheels (B and C), of which the diameters differ to the same extent as those of A and D. Since B and C are so connected that one cannot rotate without the other, their effect, when the bar is in motion, is to oblige D to revolve slightly more slowly than A, thus imparting to the screw to which D is attached a slow revolution, independent of that of the bar. Different rates of speed may be obtained by substituting differently proportioned wheels for C and D.

Another form of this arrangement, which enables the screw to be placed near the outside of the boring bar instead of its centre, will be found in Professor Goodeve's work,[1] together with another—different and somewhat simpler—method of obtaining the same motion. The machine engraved is made by Messrs. Fairbairn, Kennedy & Co., of Leeds.

In connection with the present subject we will give, before closing the chapter, one other example of boring apparatus, viz. that employed in the Royal Gun Factories at Woolwich for taking a minute final cut from the bore of heavy ordnance rifled on the system of Sir W. Armstrong. It is of course a

[1] *Elements of Mechanism*, p. 206.

tool of special application, but it is one which may be of very great value in suggesting arrangements for obtaining an equally high degree of accuracy in other cases. The boring bar for this purpose is not provided with a sliding block like those which we have just been noticing, but is held, at one extremity only, in an ordinary lathe, the gun being attached to the saddle of the slide-rest and fed up to the cutters in the manner already alluded to. The opposite or working end of the bar, which is represented in Fig. 169, is enlarged to receive the cutters. These are six in number, and are firmly screwed to it in the positions indicated by the letters A A A, being carefully adjusted as to distance from the axis of the bar by packing them at the back with very thin paper. As may be observed, they are arranged in two sets of three each, of which the first set performs almost the whole of the work—the second being chiefly added as a

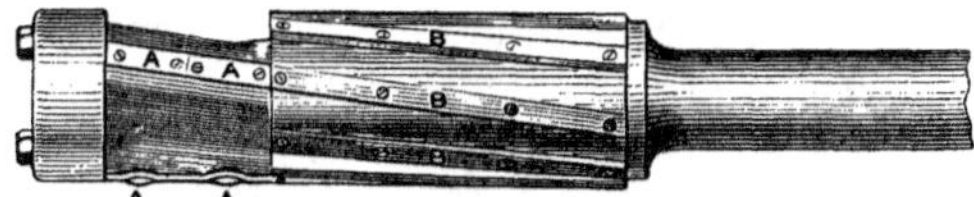

FIG. 169.—Bar for Boring Guns.

safeguard against error in the size of the bore on account of wear of the cutting edges, which takes place to a small but an appreciable extent in the course of even a single boring. Following the cutters is a series of six guide-bars (B B B), arranged spirally, which are made exactly to fit the bore. Provided that the length of these is sufficient, and their fit perfect, it is evident that the cutters cannot advance except in a straight line; so that in order to ensure the direction of the bore being correct throughout, it is only necessary for it to be correctly started. Moreover, the true circular form is effectually secured to it by the spiral arrangement of the guiding surfaces, which prevents its having any tendency to become polygonal, whilst the number of cutters in each set is that which is best adapted for ensuring steadiness in working.

CHAPTER VIII.

ON PLANING, SHAPING, AND SLOTTING MACHINERY.

ALLUSION has already been made to the important part which the 'slide principle' has played in bringing Machine-tools up to their present condition of size and power, and the reader who has glanced with us at the enormously increased efficiency of the lathe which has resulted from the introduction of the slide-rest will not require additional confirmation of the fact. But neither lathes nor drilling machines, at least in their simple forms, are dependent upon this principle to the same extent as is the class of tools which we have now to consider. Planing Machines, indeed, which in importance to the mechanic stand scarcely second to power lathes, could hardly have come into existence at all if the advantages to be derived from the use of truly plane sliding surfaces had remained unrecognised; at least, if they had, their mode of action must have been so entirely different from that in which all the varieties of those used for planing metal are made to operate at the present day, that it is difficult to conceive how they could have had a similar beneficial influence upon the practice of the workshop.

The modern *Planing Machine* is essentially a copying machine. It reduces a surface to flatness, or to a close approximation to flatness, by making upon it a series of parallel cuts with a tool which is generally more or less pointed. Each of these cuts is, as to direction, an exact copy of the movement of that portion of the machine by which the work is supported, or in other words, of the slide which determines the nature of this movement; whilst the line in which each fresh cut follows the previous one is also a copy

FIG. 170.—Planing Machine.

—in this case of the slide along which the tool receives its feed motion. If both of these movements take place in straight lines—as they must whenever they result from the sliding of plane surfaces upon one another—every portion of the work upon which the extremity of the tool has been made to act lies in one and the same plane surface; but if either the one or the other of them be made to follow a circular path, it is evident that instead of a truly flat surface we shall obtain a curved one with equal accuracy. We shall see in the course of the present chapter that this latter combination of movements, as well as the former, is turned to useful account in mechanical workshops.

A good example of an ordinary Planing Machine is given in Fig. 170. It consists essentially of a sliding table for supporting and giving a reciprocating motion to the work, together with suitable arrangements for fixing one or more planing tools above it and for providing them with an intermittent feed.

First let us examine the construction of the table, and the means adopted for imparting to it the necessary motion along the bed. The two parallel beams or cheeks which form this bed, giving it some resemblance to a lathe bed, have upon their upper surfaces V-shaped grooves, made accurately parallel throughout the entire length of the machine, and truly plane as regards their sides. Corresponding projections, which are cast on the under side of the table and are also carefully worked up to a true surface, serve to convert the bed and the table together into a gigantic slide, no other security besides the weight of the latter being necessary to keep it from leaving the grooves. The table, however, must not only be enabled, it must be compelled to slide backwards and forwards along the bed, and that both with sufficient force to cause the tool to take a cut from the surface of the work, and to a sufficiently variable distance for its motion always to correspond with the length of the portion which has to be planed. The usual method of effecting this is to cast a rack in the centre of the table on its under side, a pinion (of wrought iron) which gears into this being driven—through a train of wheels capable of sufficiently reducing the speed—by a belt and pulley in the ordinary way. Were it not for the necessity of constantly reversing the motion all would be simple enough, but after each cut has been made the table must be brought back to the position from whence it started, and since the tool cannot act when the work is moving away from its edge, it becomes an object to spend as little time as possible over the return stroke.

This has led to the introduction of various methods of ob-

FIG. 171.—Quick-Return Gearing.

taining a *quick return*, one of which will be intelligible from the accompanying diagram (Fig. 171), which shows the general arrangement of the wheels in the planing machine represented above. Their duties are threefold: First, to effect a sufficient reduction in the speed of the shafting from which the machine is driven; secondly, to convert the rotary motion thus reduced in speed into a forward horizontal one (the table as it appears in the figure being thus moved from left to right); and thirdly, to produce a similar backward movement at a less reduced speed. For the second of these—namely, the conversion of the circular into a rectilinear motion—the rack below the table, which has been already mentioned, provides. A short length of this is shown at R in the figure, the wrought iron pinion and the large spur-wheel, of which the upper portion can be seen in the general view of the machine, being represented by P and S respectively. Of the three pulleys and the remaining toothed wheels, those by which the forward stroke is given are marked A, and those for the return stroke B; the increased number of wheels in the former of these trains being for the purpose of sufficiently reducing the speed, which, as already pointed out, would be a positive disadvantage for the latter. The central pulley is 'idle,' that is to say, it runs loose upon the shaft, as also does the 'return' pulley which is marked B, but this last is inseparably connected with the pinion B. In order, therefore, to drive the table alternately backwards and forwards at these different speeds, it is only necessary to shift the strap from one of the outside pulleys to the other at proper intervals, the idle pulley between them allowing a short period of rest between the two motions. This can be done by merely moving to one side or the other the lever which projects from the bed near the centre of the machine, but by a simple arrangement, to the consideration of which we will now pass on, the table itself is made to change the position of the strap when it comes to the end of its traverse, whatever the length of that traverse may be.

Projecting from the bed of the machine near the lever just mentioned two short curved arms may be perceived, of which the resemblance to a pair of *horns* has gained for them that appellation. They also appear in the diagram (Fig. 173), where they are marked H H′, in which the portion of the figure which lies to the left of them, taken in connection with the plan of the strap-shifting arrangement shown above it, will serve to explain their action with but little verbal description. It should be noticed, however, that the horns H H′ are not in one and the same vertical plane, H′ being placed in front of H by the amount of its own thickness. By providing the table, therefore, with two adjustable *stops*, one in the plane of each of the horns, and setting these at proper distances apart for reversing the motion at any required point, the short shaft to which the horns are attached is turned to a small extent alternately to one side and to the other, thereby moving the lever L, the sliding bar G, and the fork F, between the fingers of which the strap passes to the pulley. One only of the stops appears in either the perspective view or the diagram. In the latter it is marked J.

A planing machine, however, is required to be self-acting with regard to the feed of the tool as well as in respect of its own traversing motion, and since the feed must be intermittent this too can be conveniently derived from the movement of the horns by which the shifting of the strap is accomplished. But before turning our attention to the arrangement by which the feed is made self-acting, we must examine in some detail the construction of the apparatus by which the tool is supported.

A *Tool-holder* or tool-box, as it is sometimes called, is represented in Fig. 172. The cross-slide which carries it can be seen distinctly in the general view of the machine (Fig. 170), in which also should be noticed the bevil gearing at the top of the side standards, by which the cross-slide itself can be raised or lowered. The screws within the standards, upon the upper extremities of which one of each pair of these bevil-wheels is fixed, thus form a rough adjustment for the height of the tool. But for finally adjusting it, and for giving also a vertical feed to it when required, the tool-holder is furnished with a slide, the two halves of which are marked U and V in the engraving. By turning the hand-wheel which is keyed upon the top of the vertical screw, V can be raised or lowered. The tool, however, is not attached directly to the moving half of the slide V. It is often desirable to incline it to a small extent either to one side or the other, for which purpose V has affixed to it a slewing-plate W, which is capable of being set to the required inclination, and being retained in that position by a bolt passing through a slotted hole in the upper portion of it. But even to these the tool is not directly attached. For it is found to be highly injurious to the cutting edge to keep it in contact with the work during the return stroke—an evil which is perfectly remedied by hinging the front plate (X), instead of connecting it rigidly with the slide. By this, therefore, the tool is carried.

FIG. 172.—Tool-Holder.

In addition to slewing the tool by means of the plate W, it is sometimes important to be able to give it an inclined or 'angular' feed instead of a vertical one. This is provided for by making the hinder portion of the slide (U) moveable laterally about a point somewhat below its centre, the bolts by which it is attached to the saddle T having the heads constantly in some part of a circular T-groove, so that U, and with it the entire front portion of the tool-holder, can be clamped by them at any desired inclination, the angle being indicated by graduations on the face of the saddle.

Let us now see how the feed motion can be given, whether this slide be set vertically or inclined. In Fig. 170 two light shafts may be observed running from end to end of the cross-slide, and passing behind the saddle of the tool-holder. One of them (the lower one) has a screw-thread cut upon it, by which it drives the tool-box horizontally along the cross-slide in the ordinary way. But it is the other shaft which communicates motion to the parts which we have just been examining, and this has no screw-thread, but merely a longitudinal groove. Upon it is a small bevil-wheel with a feather in its boss, which by this means is made always to revolve with the shaft, although it can travel freely with the saddle to any part of the cross-slide. Gearing into it is a second bevil-wheel, and attached to this is a third, the horizontal axis upon which these last jointly revolve coinciding with that about which the front portion of the tool-box can be adjusted. One other bevil-wheel upon the lower end of the vertical screw establishes the connection between this and the grooved shaft, the screw of course moving the slide equally whether it receives its own motion through the hand-wheel at its upper, or through the bevil-wheel at its lower extremity. The arrangement may seem to be complicated, but it appears to be the simplest which can be used for transmitting the motion between two shafts, of which the axes are in different planes, and which have also a variable inclination to one another. It enables either a vertical or an inclined (or 'angular') feed to be given to the tool with equal facility.

It now only remains for us to trace the motion to the shafts within the cross-slide, and to observe how its amount can be regulated so as to vary the feed of the tool. And by merely increasing the number of the parts the same arrangement can be made to serve when two tool-holders are attached to the cross-slide, which is generally the case in the larger sizes of planing machines. Two screwed shafts and two with grooves are then placed within the cross-slide, either pair being capable of being simultaneously set in action, so that feed motions, similar both in amount and in direction, can be imparted to both the tools. In the portion of the cross-slide which appears in Fig. 172, this duplicate set of shafts may be observed, the machine to which it was attached having been provided with two tool-holders, although—in order to avoid complication—we have selected for the general view (Fig. 170) a smaller machine which has but one.

In the diagram, Fig. 173, is the side view of a cross-slide with a single tool-holder, of which the various parts are indicated by the same letters as in the perspective view already given (Fig. 172). The ends of the horizontal screw and

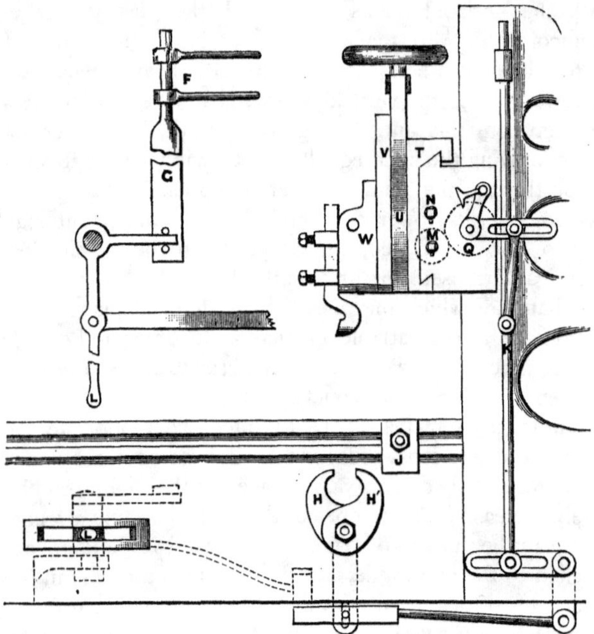

FIG. 173.—Diagram of Feed-motion and Reversing Gear.

grooved shaft appear at M and N respectively, and it is to one or the other of these that it is required to impart a small amount of rotation (in either direction) which must be capable of sufficient variation to make the feed of the tool either heavy or light. For this purpose a small spur-wheel (Q) is made to ride loose upon a short fixed spindle, which is equidistant from the shafts M and N, to either of which it can be made to transfer any motion which it receives by simply slipping a pinion over the end of the one which is to be driven by it. In the diagram, the pinion is supposed to be represented by the dotted circle round the end of the screwed shaft M, so that the result of communicating motion to the wheel Q would be to move the tool horizontally. And the manner in which the movement of the horns H H′ effects this is sufficiently obvious. The spindle which carries the spur-wheel Q, also has upon it a bell-crank lever, to one arm of which a pawl is attached. When this pawl is turned over, so that it rests upon the teeth of Q on one side of the arm or on the other, whatever rocking motion the lever receives serves to give an intermittent revolution in a corresponding direction to the spur-wheel Q. And this rocking motion is derived from the horns through a connecting-rod attached to the vertical rod K, at the foot of which is a second cranked lever. This lever—as well as the former one—has one of its arms slotted; by which means the amount of motion received, in the first place by the rod K, and secondly by the wheel Q, can be readily and accurately adjusted.

Having now a tolerably clear insight into the arrangement of the various parts in such a Planing Machine as is represented in Fig. 170, the reader who has the opportunity of examining for himself others of different construction will probably be easily able to understand their mode of action. But for his further assistance, and also for the information of those who have not that advantage, we will here briefly notice one or two of the more important of the other mechanical expedients which are to be met with in tools of this kind.

One of these—which, although the particular machine given in the above engraving is not provided with it, may be found in those of larger size by the same makers—is the *stepped rack.* In appearance, it exactly resembles an ordinary rack cut in the direction of its length into two or more equal parts, the teeth of each part being set out of line with each other—as shown in the annexed cut (Fig. 174) —to the extent of one half their pitch if the rack be divided into two parts—to one-third if into three, and so on. The pinion is, of course, made in a similar manner—the effect of the arrangement being to cause the teeth to work together with all the smoothness which would be obtained by pitching them two or three times as finely, without sacrificing the strength due to the coarser pitch. The latter, we may mention, is frequently severely tried, on account of the great weight of the table and the work carried by it in some of the larger planing machines.

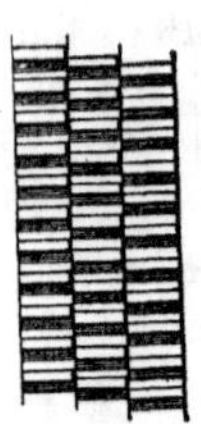

FIG. 174. Stepped Rack

But a rack is not the only method employed for converting the circular motion delivered to a machine of this kind into the intermittent one required for its table. An endless chain driven by rag-wheels has been used for the purpose, but has long been abandoned; and in the machines made at the present day by Sir Joseph Whitworth and Co., the same end is gained—in what is considered by some engineers to be a superior manner—by placing a powerful horizontal screw within the bed, the direction of its revolution being reversed at proper intervals for giving any desired length of traverse. The arrangement of the reversing gear will be evident from Fig. 175, in which, as in that already shown in Fig. 171, the motion of the table is either backwards or forwards, according as one or other of the outside ones of a set of three pulleys is driven by the strap—that in the centre being, as before, an idle pulley. The screw passes through a nut attached to the under side of the table, and has upon its extremity a bevil-wheel of sufficient size to reduce the speed of the pulleys to that required for driving the screw. Upon the opposite sides of this wheel two bevilled pinions gear into it, one of which is connected with one, and the other with the other of the two outside pulleys, by casting one pulley with its pinion upon a short hollow spindle, and passing through it the shaft to which the other pinion and pulley are keyed;—their arrangement being in this respect precisely similar to that of the pinions and the pulleys marked A and B in Fig. 171. But it is evident that although, by merely shifting the strap, the screw can then be driven alternately backwards and forwards, its speed is the same in either case,

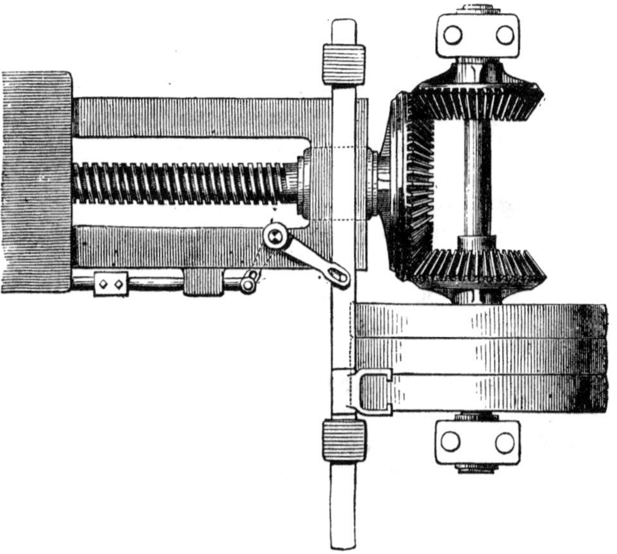

FIG. 175.—Whitworth's Reversing Gear.

so that as much time is occupied in the return as in the forward movement. This Sir Joseph Whitworth has turned to account by the simple method of reversing the tool at the end of each stroke, thereby enabling it to cut during the whole time that the table is in motion without regard to the direction in which it is going.

But whilst we are on the subject of driving and reversing

the table of planing machines, an ingenious modification of the above gearing should be mentioned. It was invented by Mr. Shanks, of Johnstone, near Glasgow, and by the addition of only one bevil-wheel it provides this arrangement with a quick return motion. The additional wheel, which is of small diameter, is fixed within and concentric with the circumference of the larger wheel which appears at the end of the screw in the preceding illustration. With this the bevilled pinion which is in connection with the reversing pulley is made to gear instead of with the large wheel, into the opposite side of which the other pinion, which runs the table forwards, gears as before. The difference between the diameters of the inner and the outer wheel carried by the screw then determines the variation in the speed of the return and the forward stroke, no alteration in the size of the pinion being necessary.

A few words will suffice to explain the general principle of the *revolving tool-holder*, which was invented, as above stated, by Sir Joseph Whitworth for the purpose of enabling the tool to cut during the return as well as during the forward stroke. In front of the holder—occupying in fact the position of the plates W and X in Fig. 172—is a vertical cylinder. At the lower end of this is a recess, in which the tool is placed, and in which it can be so adjusted that the centre of its cutting edge is exactly in a line with the axis. Round the axis the cylinder is driven through exactly half a revolution every time the table changes its direction, special arrangements being provided for giving the desired feed on each of these occasions, and not at the end of the return stroke only, as in the machine described above. This mechanism, although it appears sufficiently simple when seen in operation, would appropriate far too much of our space if we were to attempt to explain it here in detail. A description of the method by which the amount of the feed is regulated will, however, be found in Professor Goodeve's volume of this series ('Elements of Mechanism,' p. 94). The semi-revolution of the cylinder, which, with the earlier machines of this kind, was obtained from the alternate rise and fall of a stud working in a spiral groove cut upon the cylindrical surface, is now generally produced in preference by a gut band stretched over two vertical pulleys upon opposite sides of the machine, which in its passage from one to the other takes a couple of turns round the top of the cylinder of either one or both the tool-holders upon the cross-slide. But although this may be simple, in point of appearance it is hardly satisfactory for a highly-finished machine-tool, in which conspicuous cords and bands always seem rather out of place. Owing to the prominent position of the cylinder, and the alacrity with which it whisks round, a 'patent revolving tool-holder' is better known among workmen as a 'Jack-in-the-box,' or 'Jim Crow.'

A planing machine of either of the above kinds is not unfrequently provided with a detached tool-holder in addition to the one or two upon its cross-slide. This, when required, can be fixed to either of the side standards of the machine; vertical and other portions of a piece of work which it would otherwise be difficult to plane, being readily accessible to a *side-tool* supported in this manner. The feed of a side-tool is generally given by hand, but in some instances this also is made self-acting by the introduction of an additional vertical screw, so arranged as to be capable of imparting to it a sliding motion upon the standard. In other cases again, the side-tool is carried by an entirely independent standard, which is only connected with the body of the machine by transverse girders placed beneath the bed, the overhanging portions of which carry a side table upon which this separate standard is temporarily bolted. With this somewhat makeshift contrivance it becomes possible to operate upon the sides or edges of articles of which the size is too great for the machine to admit them in the ordinary way; the clear space between the side standards being under other circumstances the limiting dimension as to width, and the distance between the tool and the table when the cross-slide is fully raised, determining the height which can be admitted. But a far better way of extending these limits would appear to be a reversal of the present system—setting the tool in motion, together with the comparatively light mechanism required for its attachment and feed, and thus saving a large proportion of the power now uselessly expended in driving the table. For in planing small pieces of work, this mass, which is often very considerable, by requiring to be set in motion and again brought to rest at every stroke of the machine, consumes an altogether disproportionate amount of power; whilst in the case of large articles, which are frequently only too ponderous in themselves, it tends still further to overtax the powers of the machine.

Occasionally, indeed, this reversal is practised in a make-shift kind of way with planing machines of the ordinary construction. An 'angle chuck' of large size is affixed to the table in the manner in which articles to be planed are in general secured—namely, by passing bolts into the T-grooves with which its surface is provided—and to the vertical portion of this a horizontal slide carrying a tool-holder is attached. The work is laid alongside the bed of the machine, and is there kept stationary; the width capable of being thus treated by the tool depending upon the distance to which the horizontal slide can be made to overhang the bed. This overhanging position of the tool, however, necessarily tends to cause the table to leave the grooves, and so to produce unsteadiness of cut.

But in one variety of planing machine—the *Edge Planing Machine*, which is intended for operating on the edges of wrought-iron and other plates—the same system is carried out in a much more perfect manner. The principle of this consists in providing a fixed table of sufficient length to admit of the longest plate which has to be treated being held down upon it. This can be done—even though there may be no available holes in the plate itself—by fixing standards at the two ends of the table, and placing a girder between them containing a series of long screws, between the ends of which and the supporting surface the plate can be clipped. Parallel to the table, and running along its whole length, is a slide-face, upon which a long screw drives a carriage bearing a tool-holder, the motion of which can be reversed by a simple arrangement, at any required points. If the tool-holder be of the ordinary construction, increased speed is of course desirable during the backward journey, and by running a crossed strap upon a pulley of smaller size than that by which the screw is driven, this is easily obtained, as pointed out in another chapter (Chapter X.), but with the revolving tool-holder which Sir J. Whitworth has applied to these machines, and for which they seem to be very well adapted, no need for a quick return exists.

In this way it is easy to see that a machine could be made capable of planing any length of edge which is likely to be required, the excessive weight of a sliding-table of similar length being a strong argument against their employment for such purposes. Ordinary planing machines, however, have sometimes great length of bed, 36 feet being by no means exceptional. With a bed of this length, a cut 27 feet long can be given on any piece of work of which neither the height nor the width exceeds 10 feet. Sir Joseph Whitworth, indeed, has a machine capable of taking a single cut 40 feet in length from any article not exceeding 12 feet by 12 feet. The bed of this machine is 50 feet long—its grooves being probably the longest true surfaces which have ever been made. By the side of such a tool, that represented in the engraving (Fig. 170) would of course appear diminutive, but it is for that reason the better adapted for being brought within the very narrow compass of our page. The following are the particulars of it:—Length of bed, 14 feet; greatest length of cut, 8 feet; greatest admissible height, 3 feet; greatest admissible width, 3 feet.

For the purpose of admitting work of greater proportionate width, *Duplex Planing Machines* have been devised —two beds being fixed together side by side, each of which has an independent table. These can either be coupled together, so as to form a large single table, or they can be used separately, just as if they belonged to independent machines. These tools are sometimes provided with self-acting oiling apparatus, sufficient lubrication of the grooves on which the table slides being very important. In spite of the oil, however, the 'scored' state of the grooves in almost every large planing machine testifies to the great amount of friction which still exists between the sliding surfaces—although this might, in our opinion, be much reduced by the employment of some less fluid lubricant. But to this subject we shall have to revert in a future chapter. At the end of the bed in Fig. 170 a cistern may be observed, in which the oil is caught as it gets worked out of the grooves by the movement of the table, and from which it can be drawn off from time to time.

Passing now from large planing machines to small ones, we find other methods available for producing the motion of the table, which could not be applied when any considerable length of traverse is required. For this purpose —that is to say, for converting the rotary into rectilinear motion—some form of crank will no doubt suggest itself to the student of mechanics; and small planing machines have been made on this principle, the table being driven, by means of a connecting-rod, from a pin in the face of a horizontal disc placed beneath it. By altering the distance between this pin and the centre of the disc, the throw of the table could be adjusted. But to this arrangement there are two serious objections; in the first place, it causes one-half of the time to be wasted, through the return stroke being neither accelerated nor made use of, and secondly, there is a great variation of speed during each cut. As a remedy for these evils the following method has been employed:—A vertical disc (D, Fig. 176), having upon its face an adjustable pin (A), is so supported below the table as to form an overhanging crank. The circumference of the disc is toothed to enable it to be driven by a pinion, and in close proximity to its face is a slotted link (L), pivotted to the framing at its lower, and attached to the table by a short connecting-rod at its upper end. The effect of causing the disc to revolve in the direction indicated by the arrow, is thus to drive the table forwards powerfully, and at a slow speed, these conditions being reversed during its return.

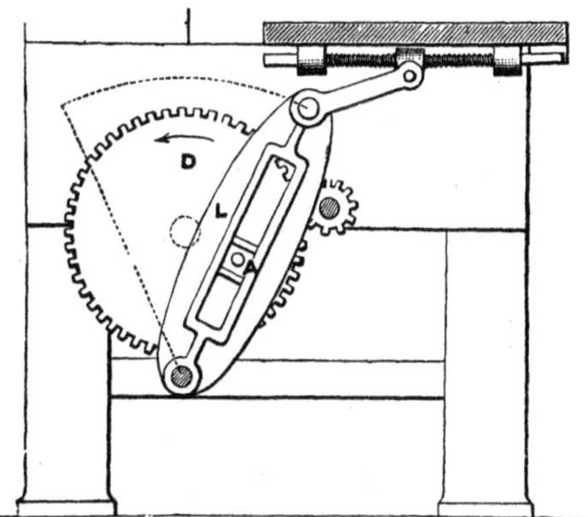

FIG. 176.—Diagram of Link Planing Machine.

But for small, if not for large work, it appears to be generally conceded that there is an advantage in giving the motion to the tool, rather than to the article to be treated by it. It is, at least, entirely on this principle that the work is performed by the instruments which we have next to consider, namely, *Shaping Machines.* Although of somewhat recent introduction, these are now amongst the most useful of engineers' tools, performing, as they do, an immense variety of minor operations, which were formerly effected by hand-chipping and filing in a much less rapid and less satisfactory manner. Fig. 177 represents one of

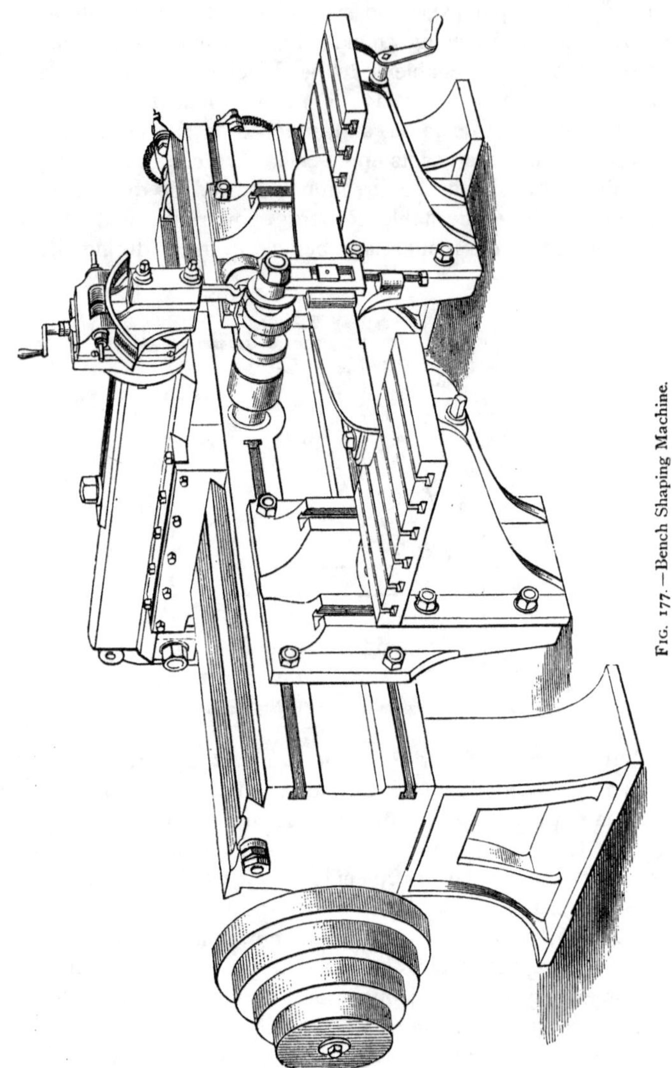

FIG. 177.—Bench Shaping Machine.

these machines, as made by Messrs. Sharp, Stewart, & Co., which may be taken as a good representative of a class, amongst the individual members of which even more than the usual amount of variation may be found.

The duties of these machines are twofold; one of them being to render flat such surfaces of moderate size as from their position, or from other causes, could not be conveniently treated in a planing machine; the other, to work up any curved faces of which the radius (for the line of curvature must form part of the circumference of a circle) does not exceed a certain fixed limit.

And in connection with the first of these we may mention —and it should perhaps have been pointed out before— that with planing machines in which the motion of the table is reversed by shifting the strap, the reversal does not take place with the exactness which is in some cases desirable, as, for instance, in planing surfaces of which the ends are in close proximity to raised portions of the work—although there is always a balance-weight, so arranged as to prevent the strap from resting upon the idle pulley, and this tends to hasten the operation. This circumstance alone frequently gives the advantage to a shaping machine, in which the length of stroke allowed to the tool can be adjusted with the greatest nicety.

In the machine shown on the opposite page, the bed bears upon its upper surface a sliding-head, capable of traversing its entire length. Upon the front of it—in addition to an arbor at the centre, with which we will not for the moment concern ourselves—there are two adjustable tables, for supporting the work and for bringing it within range of the tool. The tool-holder in which this is fixed is carried by the head at the forward end of its large and powerful cross-slide, and to the sliding portion a reciprocating motion can be given by means of the link arrangement above described (Fig. 176). In this case, however, the disc is attached to the head, and the pinion is free to travel with it to any part of a horizontal back-shaft, upon the extremity of which is the conical driving-pulley. By altering the distance between the pin (A, Fig. 176) and the centre of the disc, different lengths of throw can be given, and by means of the large hexagon nut which appears at the top of the slide—which forms the point of attachment of the connecting-rod belonging to the link—the position of the tool with respect to the front of the bed can be adjusted. With these arrangements, then, it is possible to plane up any flat surface which can be brought within reach of the tool, provided only that its width and its length do not exceed the maximum stroke of the cross-slide and the extreme traverse of the head respectively. And for this work the machine is made self-acting by merely giving an intermittent feed—derived from the back-shaft at the further end of the bed—to the screw by which motion is given to the head.

For producing curved surfaces, the work is attached to the central arbor, by being clipped between two cones in the manner shown in the engraving, or by any other convenient means. The cap of a piston-rod is there represented in position for having the curved portions of the eye planed up concentrically with the hole which runs through it; these—since they do not form a complete circle—being incapable of being turned in the lathe. Resting upon both the tables is a short girder, with an adjustment at its centre for supporting the extremity of the arbor—this, as well as the arbor itself, being easily removable when required. The curvature is produced by giving the feed-motion to the arbor, instead of to the head, the radius of the curve being ruled by the distance from its centre at which the tool is set in the first instance. The greatest diameter of work which can be thus operated upon by this machine is 12 inches, the maximum length of stroke—which equally regulates the width of surface which can be planed, whether it be flat or curved—being 13 inches. The greatest length of flat surface which it can treat is 4 feet 1 inch.

The construction of the tool-box upon the cross-slide will be obvious to the reader who has followed the description of that shown in Fig. 172, the chief point of difference between them being that in this case a tangent screw taking into the segment of a worm-wheel is substituted for the slotted hole and bolt by which inclination can be given to the tool in an ordinary planing machine. The object of these is to enable concave surfaces also to be planed, the work being kept at rest, and the feed given to the tool by means of a handle placed upon either end of the tangent screw.

The stroke of the tool is not, however, by any means always produced in the manner mentioned above. In the case of small machines with a maximum stroke of not more

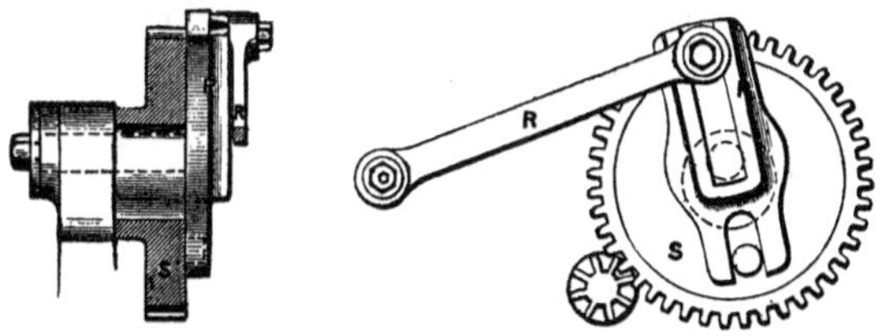

FIG. 178.—Whitworth's Quick Return for Shaping Machine.

than six inches, the additional complexity of the gearing required for giving increased speed during the return more than outweighs its advantages; but, with the exception of these, the generality of them will be found to be worked by the following ingenious arrangement, which was invented by Sir J. Whitworth for providing them with a quick return. Upon a convenient part of the head is placed a spur-wheel (S, Fig. 178) capable of revolving freely upon a fixed overhanging shaft of large diameter. Through this shaft—but not concentric with it—runs a spindle, having upon its extremity a crank-piece (P), which it keeps in contact with the face of the wheel. Thus supported, each would be able to revolve independently of the other—the wheel upon its fixed shaft, and the crank-piece round the axis of its spindle—but their revolutions would be eccentric to one another. But by slotting the crank-piece and fixing a pin in the face of the wheel, they are compelled to revolve together; the velocity of the crank—if that of the wheel be uniform—varying constantly, as the pin which drives it and the spindle which carries it, approach and recede from each other. A dovetailed groove in the face of the crank-piece, at any part of which the end of the connecting rod (R) can be set, provides the requisite adjustment for the length of the stroke.

Shaping machines are not unfrequently furnished with two heads as well as two tables, or an arbor and one table. Each part can then be used separately, and becomes to all intents and purposes a distinct tool, the bed alone being common to the two. A vertical face with T-grooves in it is sometimes a useful addition to the table.

The tools used in planing and in shaping machines resemble one another so closely, that we have purposely deferred the few lines we have to devote to the former till the latter could also be included. In planing metal, as already mentioned, the general truth of the surface produced depends upon the accurate guidance of the work and of the tool, and not upon the form of the edge by which the individual cuts are made. This being so, a pointed, or but slightly-rounded tool is almost always employed, the preference being given to these—as in metal turning—on account of their combining great strength of form with facility in cutting. A series of grooves is the result, of which the curvature of the extremity of the tool and the amount of feed determine the depth and the distance apart, the slight ridges between them being immaterial for many purposes. But when general flatness has been obtained in this manner, greater smoothness can be easily given to a surface, when required, by going over it a second time with a broad-edged tool which removes the ridges. Mr. W. F. Smith, of Manchester, has shown that greatly increased smoothness may be obtained in the first instance by using a tool with a loose cutter of circular section instead of the ordinary pointed tool. The diagram (Fig. 179) shows the effect of each, the vertical

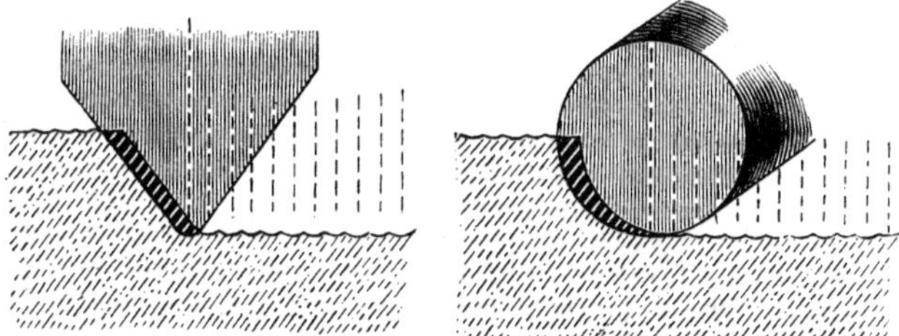

FIG. 179.—Diagram of Sections of Planed Surfaces.

dotted lines indicating the amount of feed—nine cuts to the inch in each case. A drawing of one of these cutters with its holder has been given (in Fig. 151) amongst the tools for turning metal, its form being well adapted for the treatment of cast or wrought iron, either in the lathe or the planing machine.

Much resemblance may also be noticed between the turning-tools in Fig. 150, and those ordinarily used for planing and shaping which are given below (Fig. 180). Of these the first, marked *a*, is the usual tool for surfacing, its point being slightly rounded off, as appears in the preceding diagram. For smoothing, a tool resembling this, but having a broad straight edge instead of a point is employed. Of the others, *d* is for under-cutting, and *b* and *c* are two kinds of parting tool which are used for making

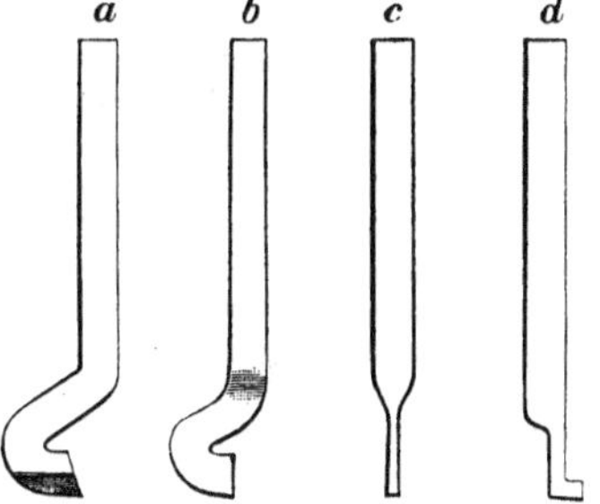

FIG. 180.—Planing and Shaping Tools.

through cuts of as small width as possible, so that their edges are kept as narrow as is consistent with their own safety. In shaping machines, right and left side-tools, similar to the turning-tools of the same name, are also sometimes employed. The cutting edge should be made in a line with the back of the shank, see *a b* Fig. 180, so that the tool may spring over rather than dig into the work.

One other machine-tool requires to be briefly noticed before we conclude the present chapter,—namely, the *Slotting Machine.* Although in principle it differs but little from a shaping machine—the slide which gives motion to the tool being merely placed vertically instead of in a horizontal position, much practical difference will be found to exist between them. Neither in their proportions, indeed, nor in general appearance do these two classes of instruments possess much in common, those which we have just quitted being some of the lightest, whereas many of these are amongst the heaviest of ordinary workshop cutting-tools. To the much more arduous nature of their duties this fact bears testimony—that their driving apparatus is almost invariably either double- or treble-geared, the latter being by no means uncommon.

The slotting or vertical shaping machine represented in Fig. 181, has gearing of the former kind, its framing and the table below its slide for the support of the work having much resemblance to those of a drilling machine. Its capabilities are therefore estimated in a very similar manner, the maximum stroke of the slide, and the space between the point of the tool and the nearest vertical portion of the framing, ruling respectively the length of cut and the distance from the edge of the work at which it can be made. In the present instance, the greatest possible length of stroke is 13 inches, and the largest diameter of work in any part of which a cut can be given, is 4 feet 6 inches—half of this (2 feet 3 inches) being therefore the distance between the tool and the framing.

But a slotting machine must be able to accommodate itself to the performance of duties of a very varying character, for besides the 'slotting' (or perhaps more properly the slitting) of circular holes so as to give them the elongated

FIG. 181.—Vertical Shaping or Slotting Machine.

section from which these tools have obtained their name—which section, as we have seen, can now be produced by other means—they are employed for almost all the paring and shaping which is required in the treatment of work of a heavy description. Ready adjustability of the height of the sliding-bar and of the length of its stroke is therefore essential, whilst a quick return is as desirable in this as in the other cases of single-acting tools which we have been examining. Let us see what provision is made to supply these several wants, and also to impart to the table the requisite feed-motion, for it is evident that the construction in the present case does not admit of this being given to the tool.

At the back of the vertical slide is a disc (D) having in its face a radial groove, in any part of which, by means of an adjusting-screw, a projecting stud can be set. Upon the amount of the eccentricity of this stud the length of the stroke of the machine depends, its motion being transmitted by a connecting-rod (C). This, however, is not attached to any fixed portion of the sliding-bar, but to a pin which can be moved upwards or downwards—also by means of an adjusting-screw—in a long slotted hole in the upper part of it. The hand-wheel at the end of this adjusting-screw appears in the engraving at the upper extremity of the sliding-bar, which can thus be readily raised or lowered.

But how, it may be asked, can the speed be rendered faster in the upward than in the downward stroke on this system, for if the revolution of the disc be uniform, the sliding-bar must occupy equal spaces of time in its ascent and descent? We answer that, in the machine shown in the engraving, the disc is not driven at a uniform speed. It receives its motion through the wheels marked E E′, and these being elliptical instead of circular so vary its velocity, that the speed which it imparts to the sliding-bar is slow and comparatively uniform, whilst the tool is cutting, and of greatly increased rapidity during the return stroke.

We have now only to examine the construction of the table and the arrangement by which its movements are made self-acting. In doing this we shall derive much assistance from the subjoined figure (182), which shows in outline the opposite side of a self-acting table of a slotting machine by

Messrs. Fairbairn, Kennedy, Naylor & Co., which is of somewhat larger size than that in the general view given above.

The table is capable of receiving three separate horizontal movements; a slide being so arranged that it can approach or recede from the standard, a cross-slide providing for its travelling in a direction at right angles to this, and a pivot at its centre enabling it to revolve about its axis. The two slides—which together form what is termed a 'compound slide'—are worked by screws in the ordinary way, and the table is turned by a worm taking into a worm-wheel immediately below it. Ratchet wheels with double pawls are placed upon the ends of the two screws, and upon the shaft which

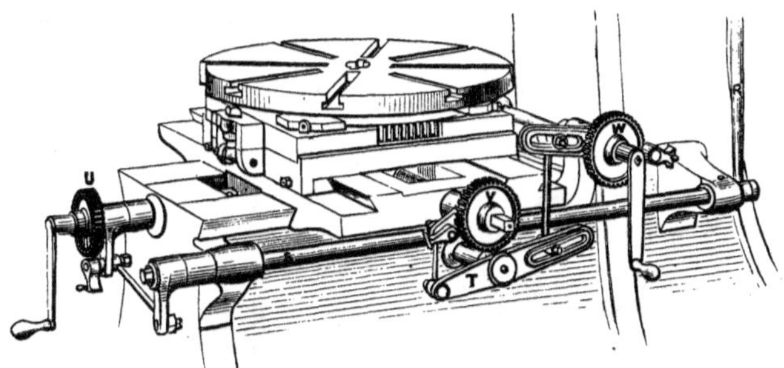

FIG. 182.—Self-acting Gearing for Slotting Machine.

carries the worm, so that any one of these can be made to receive an intermittent feed-motion in either direction by giving a reciprocating movement to the arms to which the pawls are attached. This movement they constantly receive from the shaft S, which is driven backwards and forwards through a portion of a revolution by the vertical rod R whenever the machine is in motion. R, in turn, derives its motion from a cam-groove on some convenient part of the driving-shaft of the tool-slide, its upper end being attached to a slotted link, by which the amount of its rise and fall—and consequently of the feed of the table—can be regulated. In Fig. 181 a portion of the cam-groove can be seen on the face of the wheel E′. The manner in which the motion is imparted by the shaft S to the ratchet wheel U is sufficiently evident, and the expedient by which one of a pair of bevil-wheels can be made to slide to any part of its shaft whilst still sharing its rotation—which is here employed for giving a rocking motion to the lever T from the shaft S—must now be too familiar to the reader to need further description. In addition to these motions the tables of slotting machines are occasionally provided with apparatus for enabling other curves to be cut. Provision is also sometimes made for tilting the table instead of having its surface always horizontal.

With respect to the tools for these machines, it is evident that some modification of the forms employed for planing and shaping is required to meet the altered conditions under which they are used; one main point of difference being in the direction of the thrust upon them, which is almost always parallel or nearly parallel to the shank instead of at right angles to it. Further it is frequently necessary for a slotting tool to be able to work in a very limited space; for instance, in cutting a key-bed in the boss of a wheel or in commencing a cut from any drilled hole. For such purposes a *Parting-tool* resembling that shown on the left-hand side in Fig. 183 must be employed, but in most other cases the form of tool on the right is more convenient. This is called a *Round-edge*, a *Diamond-point*, or a *Keyway-tool* according to the shape of its cutting edge, the edge of the parting-tool being also rounded when occasion requires it. Where, however, there is practically no limit to the space occupied by the tool, its position is occasionally reversed, the shank being

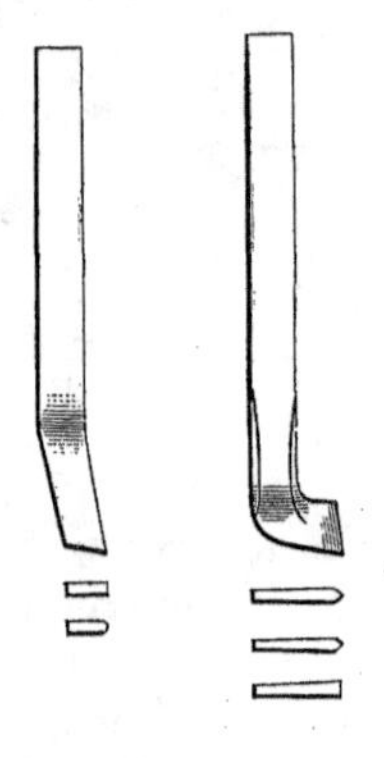

FIG. 183.
Tools for Slotting Machine.

placed at the bottom of the tool-slide instead of in front of it. The last of the three tools in Fig. 183 is intended to be used in this manner, and is called a *Right-angle* or Hook-tool.

We have now noticed the more important of the ordinary machine-tools which are to be met with in mechanical workshops—such of them at least as perform their operations by what we have throughout considered to be true cutting processes—and our sketch, although a mere outline, must here be brought to an end. It must not, indeed, be imagined that those which have been selected for our engravings are the only forms which are given to these tools, nor even that they are necessarily in every respect the best; but since full descriptions of the peculiarities of construction employed by different makers or for special kinds of work could not be attempted, we have preferred for the most part to confine our attention to a single good and recent example in each case. In this manner we had been in hopes of being able to include various other machine cutting-tools, which may be regarded rather as the luxuries than as the necessaries of a workshop, but which are, nevertheless, of considerable interest. Our rapidly diminishing space must, however, be devoted to other subjects—a circumstance we much regret, since it will entail the exclusion of the band and other sawing machines which are now used with such wonderful effect upon cold metal, the screwing and shaping machines for bolts and nuts, together with various others.

But in connection with machine cutting-tools one subject must not be passed over—namely, the means adopted for maintaining in an efficient state the cutting edges by which the work is actually performed. This, when once the appropriate form and hardness have been given by forging and tempering—concerning the latter of which we have had something to say in a previous chapter (Chap. III.)—resolves itself merely into judiciously applying that most important adjunct, the grindstone. 'Show me the grindstone, and I will tell you the character of the work which can be turned out in the shop,' was the forcible remark of one of our most eminent engine-builders; and if the mention of this should be the means of calling attention to the great advantage which would in many cases result from improvement in its condition, good service will have been done. For with a bad or an ill-used grindstone, tools cannot be properly ground, and with irregularly ground tools uniformly good work is out of the question.

First with respect to the method of hanging a grindstone, which for a machine shop is generally from 2 to 5 feet in

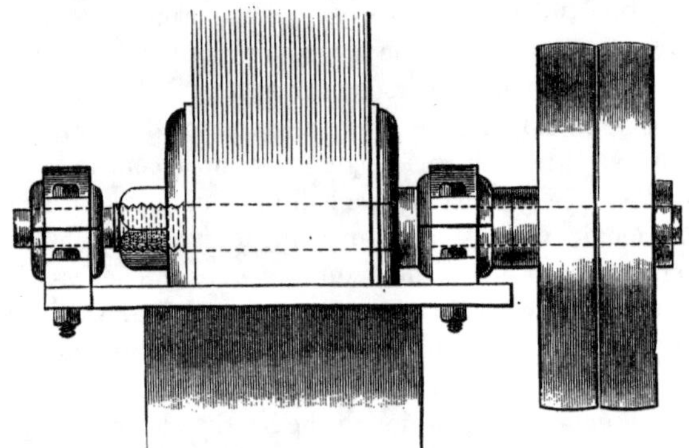

FIG. 184.—Method of Hanging Grindstone.

diameter, and from 4 to 8 inches in thickness. The best system, undoubtedly, is that shown in Fig. 184, in which the stone is firmly clipped between two circular cast-iron plates upon the spindle. Between each of them and the side of the stone a thin piece of soft wood 'packing' is placed; which accommodates itself to any slight irregularities of the surface and gives the plates a firmer grasp. The spindle runs in two plummer-blocks fixed on the sides of the cast-iron water trough, one end of it being provided with an ordinary fast and loose pulley. The trough also carries at one end a wrought-iron rest adjustable as to height, with the assistance of which, as long as the stone runs true, tools can be ground by a practised hand in a very efficient manner.

FIG. 185.—Grindstone with Slide-Rest.

But no sooner does the grindstone lose its circular form thân the rest becomes worse than useless; and this is always liable to happen, either from careless treatment, or through one part being softened by long standing in the water.

This evil can only be remedied, and indeed, its increase can only be prevented, by turning the face of the stone.

But even when the condition of the grindstone is perfectly satisfactory, the correctness of the angles which the facets of a cutting edge ground upon it form with each other and with the shank of the tool, depends entirely upon the skill of the operator, although of course he may, and in general he should, have the assistance of a gauge or template. But either with this or without it, very great advantages will almost always be found to accrue from the adoption of the system of Machine Grinding introduced not long since by Sir Joseph Whitworth. Fig. 185 shows one form of his apparatus for the purpose, which is in fact an application of the slide principle to the grinding of tools. Instead of the fixed rest mentioned above, the end of the trough is made to carry a slide-rest, of which the cross-slide is driven by a screw in the ordinary way, and the lower slide by a rack and pinion. By working the handle attached to the pinion, the tool (when its extremity is being ground) can be easily kept in motion, so that all scoring and grooving of the face of the stone are prevented. The required angle can be given with unfailing certainty by a single adjustment of the rest, the entire body of which is hinged to the trough for this purpose. Perfect uniformity in the angles of any number of tools can thus be obtained, and they can be ground not only more rapidly, but in a very superior manner. Moreover, all unequal wear of the stone is prevented, so that the process of turning it up need never be resorted to. The speed at which the face of the stone is driven is about 830 feet per minute.

CHAPTER IX.

ON PUNCHING AND SHEARING MACHINERY.

IN a former chapter, when discussing the peculiarities of some of the hand tools ordinarily used for metal, we mentioned that certain kinds perform their duties by a species of tearing, rather than a true cutting action. These are chiefly known as 'punches' and 'shears,' and the principle upon which they act is by no means confined to tools used by hand; the extension of it into the province of machine-tools providing us with the means of producing some of the most startling results which are to be seen in mechanical workshops. Thus shears, of which the small representatives worked by hand can only with great difficulty cut through sheet iron ⅛th of an inch in thickness, will, when fixed in a suitable machine, pass in a very small fraction of a minute through as much as an inch and a half of solid plate, or will snip off a 6-inch bar of nearly double that thickness in a similar space of time, with, to all appearance, the greatest ease imaginable.

With these Punches—which equally perform feats in their own line which, but a few years since, were beyond the highest aspirations of the millwright—are, more often than not, combined in one and the same machine,—so that, independently of the closely analogous manner in which they act, it will be advisable, even if not absolutely necessary, to describe Punching and Shearing Machines together.

It will be remembered that before we quitted the subject of hand-punching, one or two instruments were mentioned which, although worked by hand, were capable of perforating plates of considerable thickness. Such are the 'punching bear' and other portable forms of apparatus, which, by the use of simple mechanical expedients, enable these results to be obtained by the continued application of so small a force as that which can be exerted by the human arm. In addition to these, there are also other machines of various degrees of simplicity and portability,

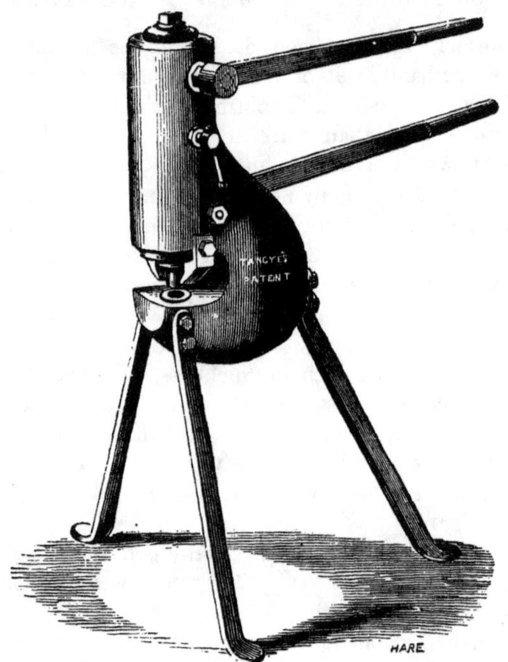

FIG. 186.—Hydraulic Bear.

which, since their effects depend solely upon long-continued or accumulated hand-power, might perhaps have been more properly included in a former chapter. Foremost among them stand those which are worked by hydraulic pressure —the almost perfect incompressibility of water affording us the means of concentrating a succession of small efforts in a more perfect manner than can be done by any mechanical expedient with which we are acquainted. Fig. 186 represents a *Hydraulic Bear*, constructed by Messrs. Tangye, Brothers, with which holes of as much as 1¼ inch in diameter can be punched in wrought iron 1 inch in thickness by the power of only one man. Its form will be seen to bear a general resemblance to that of the screw-punching bear (Fig. 81), with the exception of the legs with which this is provided, which are not an essential part of it, and can be easily detached when necessary. Into the action of the hydraulic press, of which this is merely a modification, we cannot here enter fully; but the general construction of this tool will be easily understood from Fig. 187. In this, which is a vertical section through it, A is a cistern containing a supply of water, C is the plunger of the force-pump D, and B the lever by which it is worked. E is the inlet-valve by which water enters the pump when the plunger is raised, an outlet-valve at the foot of this allowing it to pass into the cylinder F on the descent of the plunger. In this cylinder is the ram, H, which has a cupped leather packing at its upper, and the punch at its lower end. L is a lever for raising the ram, and so withdrawing the punch from the hole which has been made by it. The student will be aware that (friction neglected) the pressure exerted by the ram exceeds that applied to the plunger, inversely in the proportion which the area of the latter bears to that of the former.

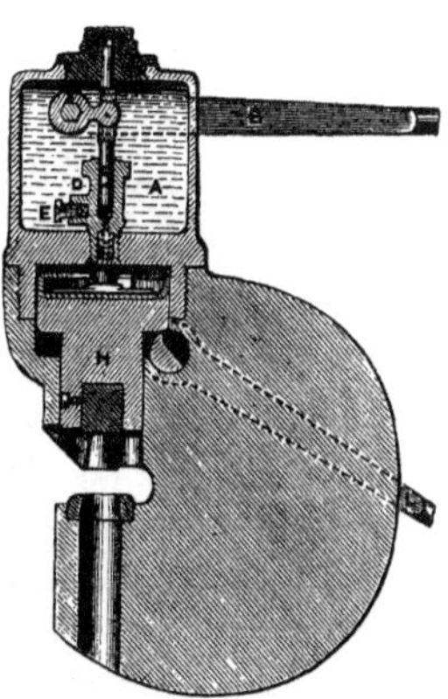

FIG. 187.—Section of Hydraulic Bear.

Besides the punching bear given in Fig. 186, other forms of hydraulic punch, both portable and stationary, are to be met with. When an 'open mouth' is unnecessary, as in punching the webs of rails, &c., a more advantageous disposition of the metal of the framing in the neighbourhood of the punch becomes possible, the vertical portion being made in two parts, and the rail or bar which is to be

FIG. 188.—Hydraulic Shears.

punched being passed between them. This arrangement is known as the 'close mouth,' and is that with which the *Hydraulic Shears*, Fig. 188, are provided. The action of this tool being in all its main points similar to that of the punching bear just described, will not require special explanation. The lever for raising the ram is omitted in the figure, its position, when in use, being on the spindle A. During this operation, the screw immediately below the pump-lever is slackened, a stop-valve in connection with it then allowing the water in the cylinder to pass back into the cistern.

But the operation of punching may be performed in a much more rapid manner by means of a *Fly Press*, which, at first sight, might be thought actually to give off more power than had been applied to it—a result which every student of mechanics knows to be an absolute impossibility. One of these instruments may be occasionally seen, mounted on a 'trolly,' by the side of our lines of railway, with which two men are able to punch a hole nearly an inch in diameter through the 'web' of a rail, perhaps three-quarters of an inch in thickness. The explanation of its action lies in its power of accumulating the force applied to it, which, although in itself it may be small, is continued for a sufficient time to produce these apparently disproportionate results. The fly—which generally consists of two ponderous balls of metal at the extremities of the lever of the press—is gradually brought from a state of rest to one of rapid motion by continued effort on the part of the workmen—the maintenance of its speed being assisted by its descent due to the revolution of the screw. But, inasmuch as it parts with the whole of its momentum almost instantaneously on the punch coming into contact with the rail, the results are proportionate to the force and the length of time expended in bringing the fly to its full speed, and not to the duration of its performance. But here we are trenching upon ground which lies more properly within the province of another forthcoming volume of this series.[1]

In stamping coins, and some other processes which are rather beyond our limits, fly presses of much greater size and power are employed, the recoil—when the motion is quite suddenly arrested, as it frequently is in such cases—being sufficient to raise the screw to its original position.

[1] *Principles of Mechanism:* By T. M. Goodeve, M.A.

Passing now to the stationary forms of punching and shearing machines, which belong more properly to our subject, since in some form or other they are to be found in every mechanical workshop; one which is almost, if not absolutely, the simplest, is that of which an outline diagram is given in Fig. 189. Although one of the earliest, it possesses certain advantages over the somewhat neater arrangements presently to be noticed, which have caused it to be generally preferred by boiler makers and others, to whom an efficient punching machine is an essential. Its

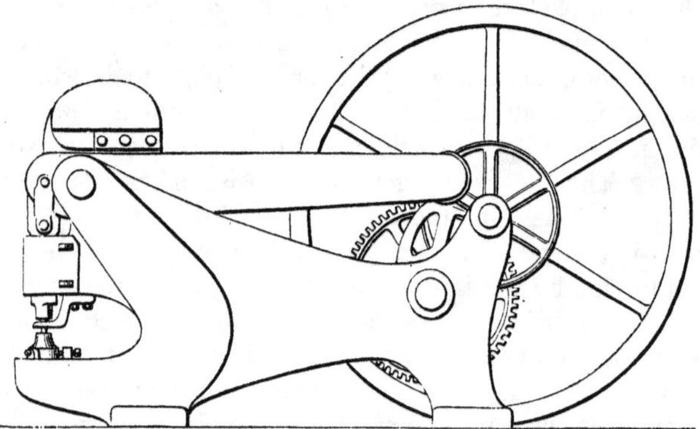

FIG. 189.—Lever Punching Machine.

construction will be easily understood from the engraving. The punch is attached to the lower end of a guide-bar, which is connected with the shorter arm of a powerful lever, so as to rise and fall with it. The longer arm of this lever—which carries a roller at its extremity—rests upon the edge of a slowly revolving cam—its motion being derived from the fly-wheel shaft by means of a spur-wheel and small pinion. It will thus be seen that the revolution of the cam, and, consequently, the motion of the lever, cannot be arrested without also bringing the fly-wheel to rest. Its momentum is therefore available for overcoming any resistance which the punch may encounter. After each descent the punch is raised by the mere weight of the longer arm of the lever. The following advantages accrue from regulating the motions of the punch by a cam, instead of by an eccentric, as in the machines described below. In the first place, it follows that the descent of the punch can be made perfectly uniform; 2ndly, that its return can be accelerated to any required extent; and, 3rdly, that by causing it to remain perfectly stationary between the end of one stroke and the commencement of the next—for which it is merely necessary for a portion of the cam to be made concentric—ample time can be given for the workman to place the plate correctly.

We have already called attention to the hollow cast-iron frames which are now generally to be met with in machine-tools. In the case of punching machines they are required to be of unusual strength to resist the severe strain due to the passage of the punch through the plate. This is secured by forming at the working end of the frame a powerful fixed jaw, to the upper part of which the lever is centred, the bolster or *die* being placed upon the lower part. The distance between the inside of this jaw and the centre of the punch is an important factor in determining the useful qualities of these machines, since it limits the distance from the edge of a plate at which any hole can be punched by them. But every increase in this distance requires that much more than an equal increment of strength be given to the jaw.

Above the lever, at a short distance from its fulcrum, a pair of knives may be observed in Fig. 189, an addition which enables this machine to be used also for shearing when required. Their position, however, is by no means a convenient one, the most advantageous height for them being that occupied in this instance by the punch and die. Some machines of this kind, indeed, are made for shearing

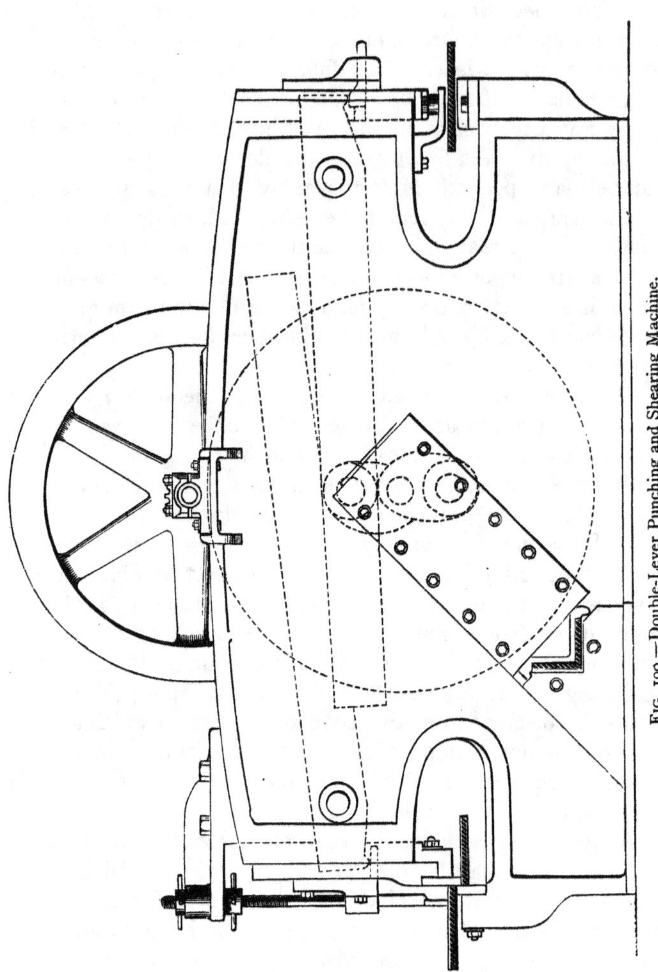

Fig. 190.—Double-Lever Punching and Shearing Machine.

only, in which case knives are substituted for the punch and die, the upper one being worked by the lever and cam in the manner shown in the diagram; and it must be confessed that the arguments in favour of this method of driving are as forcible for shearing as for punching.

A combined punching and shearing machine, made by Messrs. Collier of Manchester, is shown in outline in Fig. 190, which, whilst it possesses much of the compactness of appearance of those on the eccentric system next to be described, retains the advantages of the cams and levers already mentioned. The shears are placed on the same level as the punch, but at the opposite extremity of the machine, so that the framing has in this case a jaw at each end, the two independent levers by which they are worked being within the hollow of the framing. The cams by which motion is given to the levers are fixed upon the spindle of a large-sized spur-wheel (shown by dotted lines in the engraving), which is driven by a small pinion on the fly-wheel shaft; and since the greatest radius of the one cam is on the same side of the shaft as the smallest radius of the other, each lever is raised alternately. By this arrangement it becomes impossible to put an undue strain upon the machine by attempting to use it for both punching and shearing at the same moment. It may be observed that there are several points of difference between the cam and lever of the single-ended machine and those of the combined machine which we are now discussing. In the present instance the cams do not act upon the extremities of the longer arms of the levers—which they could not do without considerable addition to the total length of the machine. Consequently the power obtained is proportionately reduced, although in many cases the much greater economy of space which results from making both ends of the machine available, may more than compensate for the loss of it. For in spite of all that may be said in favour of the single-lever machines it cannot be denied that they monopolise an amount of space which generally makes them inadmissible in a crowded workshop.

In Fig. 191, which represents a partial section of the punching end of one of these combined machines, some further points of difference may be noticed—amongst others the transference of the friction roller from the long arm of the lever E to the portion B of the cam C, at which the greatest pressure is exerted, and the insertion of the nose or

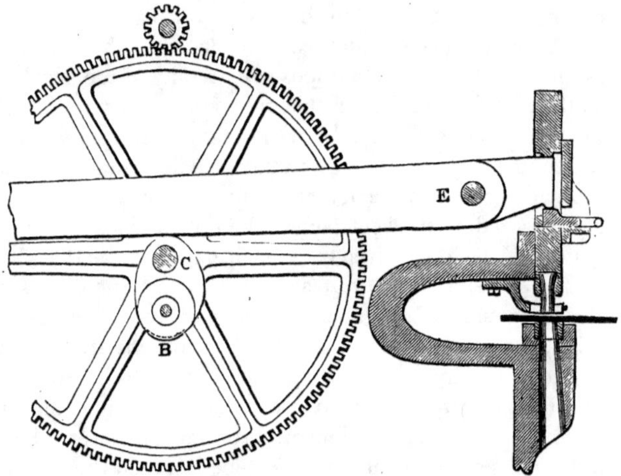

FIG. 191.—Section of Punching End of Double-Lever Machine.

short arm into a recess in the slide which carries the punch, instead of their connection being established by separate links. One great advantage of this is the facility which it affords for throwing the punch out of gear, in the manner explained below. To ensure the weight of the lever being competent to withdraw the punch and to raise the slide, its length is made much more than sufficient to reach the cam at the centre of the machine. One other portion of this tool demands a few words of explanation, viz., the projection which appears in Fig. 190, extending downwards from the centre at an angle of 45°. Near its lower extremity an L-shaped opening may be noticed. In this is a pair of shears adapted for cutting bars of angle iron, the upper knife being driven by an eccentric at the end of the shaft which carries the spur-wheel and cams. But that this method of driving is inferior will be shown in connection with the punching and shearing machine, which we have next to describe. For cutting other sections of rolled iron various other forms of knife are employed.

Fig. 192 is a general view of a combined *Punching and Shearing Machine*, by Messrs. Sharp, Stewart and Co., in which both the shears and the punch are worked by eccentrics; the compact form which can be given to the framing of machines of this kind contrasting very favourably with those of the lever punches and shears which we have been discussing. Within the hollow framing is a strong horizontal shaft, upon which is fixed the large and powerful spur-wheel which appears in the centre of the figure. At the extremities of this shaft are eccentrics connected with the upper knife of the shears and the punch respectively, one of them always rising during the descent of the other. The manner in which the vertical movement of the slides to which the punch and the knife are attached, is obtained from these eccentrics is slightly dissimilar, owing to the difference in their respective requirements. The shear-blade is attached to the pin of the eccentric by means of a link (the upper joint of which can be seen in the engraving), so that it rises and falls at every revolution of the shaft, and can never be disconnected from it. In the case of shears, indeed, this is never required, although for a punch it is a great desideratum. A slight modification at the punching end of the machine enables the connection to be instantly broken and as quickly re-established. The link is not perforated at both ends, as before, but the lower end is made to drive the punch by simply pressing downwards upon a projecting portion at the centre of the slide, with which it only comes into contact when its lower extremity is central. A handle at the end of the machine enables the link to be turned aside so as to miss the pro-

FIG. 192.—Double-Geared Punching and Shearing Machine.

jection during its descent, so that the punch remains at the top of its traverse—a weighted lever, working on the centre which appears in the figure above the jaw, continually retaining it in that position when it is disconnected from the link.

A lever punch, however, admits of still more simple means being employed for this purpose. The opening in the slide, into which the nose of the lever is inserted, is made of sufficient length to admit not only the lever, but also a loose piece, or 'block,' which can be slipped in below the lever through an opening in the framing. This block appears in both the outline elevation (Fig. 190) and the section (Fig. 191). On the punch in its descent coming into contact with the work, the block, when in position, transmits the downward pressure of the lever to the slide, whence it passes to the punch. When, however, the block is withdrawn, the slide merely falls by its own weight, the punch descending and resting upon the work till it is again raised by the lever.

But two other points remain to be noticed in Fig. 192, although they must not be supposed to be peculiar to this class of machine. One of these is the vertical plane of the shear-knives, which is not at right angles, but greatly inclined to the sides of the framing, and to the axis of the eccentric shaft. This is done with the object of enabling them to operate upon bars or plates of any length, provided only that they are of small width. If placed square with the framing, the depth of the jaw would of course limit the distance from the end or edge of a plate at which the shears could operate, as already pointed out with reference to punching. The other point is, that the fit of the slides—on the freedom and steadiness of whose motion so much depends—can be adjusted by means of setting-up screws, five of which appear in the engraving at each end of the machine.

Some single-ended machines are also constructed on the eccentric principle, a single slide being then made to do double duty, having the punch attached to its lower, and the bottom shear-knife to its upper end. For workshops in which economy of space is a greater object than convenience of position for the shears, this arrangement is not easily improved upon—the eccentric having, as has already been pointed out, undeniable advantages in enabling the length of the framing to be reduced. Where, however, there is sufficient space to admit of the use of lever machines, it is not a little surprising that those driven by eccentrics should so frequently be found. Briefly, the requirements of a punching and, to some extent also, of a shearing machine, are that the motion of the punch or knife shall be slow and uniform whilst cutting, quick during its return, and that the interval of rest shall bear as large a proportion as possible to the time occupied by the stroke. These, as we have already stated, are well fulfilled by a properly-shaped cam and lever, whereas the eccentric gives us variable and similar motions during both the upward and the downward stroke, and commences its return immediately on arriving at the end of its traverse. The last, indeed, is a much more weighty objection to this application of the eccentric than might at first be supposed, since the accuracy of the position of holes punched in the ordinary manner (which are merely set out with a template by a kind of stencilling process) depends solely upon the plate being correctly placed by the workman, and the shorter the interval at his disposal for the purpose, the greater is the risk of error. The superiority of well-constructed lever machines in this respect may be gathered from the fact that on this system it has been found possible to keep the punch absolutely at rest during one-half of each revolution of the cam, thereby enabling the number of strokes in a given time to be made one-third greater than would be attainable in an eccentric machine, without curtailing the interval for placing the work.

To obtain more perfect correspondence between the holes in any two or more plates than is possible when the marks which denote their position are brought separately under the punch by hand, Mr. Maudslay many years ago introduced a system of *Multiple-Punching*. The machines for this purpose, some of which have remained in use to the present time, bore a general resemblance to Fig. 190, the required motion, however, being imparted to the lever by a pair of toothed elliptical wheels, one of which occupied the position of the cam, and the other that of the roller. A series of punches were arranged in line at the lower end of the vertical slide, excessive strain upon the machine being prevented by making them of different lengths, so that they acted in succession, instead of at the same moment. A self-acting 'dividing table,' by which the plate was moved through the correct distance after each stroke of the machine, enabled greatly increased accuracy of 'pitch' (or distance from centre to centre) to be given to any series of holes. Single punching machines with self-acting tables are now made by Messrs. Fairbairn and Co., and by other makers, and they appear to have much to recommend them. Other machines, with still greater capabilities and of correspondingly greater complication, have been devised, in some of which several rows of holes of equal pitch can be punched simultaneously; in others—or at least in one, the 'Jacquard Machine,' invented by Messrs. Roberts of Manchester—any arbitrary arrangement of holes can be produced. But these—chiefly on account of their excessive cost—are rarely to be met with, and therefore do not demand special notice here.

Recently, indeed, the tendency has been to diminish rather than to extend the application of punching machines, and to substitute drilled holes in many cases where they would formerly have been punched. As yet, however, even multiple drilling has not been put on a par with punching in respect of speed—from 8 to 10 holes per minute being capable of being punched singly even with machines of the largest description, and as many as 15 being possible with those of more moderate size, which more than equals the performance of the multiple-drill, already noticed.

But it must be remembered that we have throughout the

present chapter been speaking only of the machines used by makers of boilers and iron structures for punching wrought-iron plates, bars, &c., which form but one branch of a very wide subject. Into its other ramifications our subject does not carry us, although some of them possess great mechanical interest—as, for example, the process of perforating zinc, &c., with Larivière's machine, by which three rows of perforations, containing some hundreds of holes, are made simultaneously. For almost all purposes, indeed, for which it is necessary at a single operation to make holes which are otherwise than circular, recourse must be had to punches and dies, with which those of the most complex forms can be produced with nearly as much facility as can simple circles, ovals, or squares. To this, however, we have seen that the process of slot-drilling forms a notable exception.

The fact of the metal removed by a punch and die being in one solid piece or 'burr,' instead of being frittered away in shavings, is also turned to useful account in numberless manufacturing processes—'blanks' for coins, buttons, percussion caps, steel pens, the wheels and other portions of cheap clocks, and an endless variety of small metal objects of daily occurrence being made in this way. For some purposes, indeed, articles of much larger size are reduced to their proper shape in this manner. The axle-guards for railway carriages are a case in point, which, although they measure some 2 feet in height and exceed a foot in width, are cut at a single stroke from iron plates $\frac{3}{4}$ of an inch thick by a species of punching or shearing machine—for in such cases it becomes difficult to say under which head it should be included. But such processes are not usually in much favour in mechanical workshops. Still less are those in which a single hollowed cutting-punch is used, instead of the solid one with a corresponding die, which is the form employed in all the machines which we have been describing. The others, indeed, are chiefly used for soft materials ; but a modification of them, in which two corresponding punches, with strong cutting edges, are forcibly driven from opposite sides into a sheet or thin plate of metal, has been of late years gaining ground in the manufacture of small articles, though we are not aware that it has yet extended into the practice of the engineer.

Before quitting the subject of punching machinery, one or two points in connection with the aforesaid dies and punches should be briefly noticed, inasmuch as the quality of the work performed by any punching machine is in the highest degree dependent upon the form and proportions of the punch and die. The use of the latter was explained in a former chapter (Chap. III.), and its general appearance and the means of adjusting it, so as to ensure its being in every instance concentric with the punch, will probably have been observed in the preceding engravings, although the attention of the reader has not before been directly called to it. In the hydraulic bear, for instance, a cast-steel ring forms the die (or 'bolster' as it is not unfrequently called in the case of the smaller punching instruments), and it is kept concentric with the punch by being turned on the outside, so as to fit into a recess in the cast-iron jaw. But in the larger machines it is found more convenient to widen out the base of the die, and merely to let it rest upon the flat surface of the jaw, its position being regulated by three or four screws working in fixed lugs, or by some other simple adjustment. Some makers transfer this adjustment to a separate bolster, into the upper part of which a small die is fitted, an arrangement which does not necessitate so great an expenditure of steel. But, whatever may be the external form of the die, a central perforation extends completely through it in all cases, and it is upon the internal proportions of this that its efficiency depends. The upper part of the aperture is cylindrical ; the lower being tapered, as the accompanying section of a die and bolster (Fig. 193) shows. The parallel portion can then be sharpened by grinding its upper surface, as the angles become rounded by use, without altering the size of the hole, whilst the tapered continuation of the hole allows the burr which is punched out to fall freely through by its own weight—an opening being cast in the jaw just below the die, so that it drops out altogether clear of the machine. For the actual size of the cylindrical portion as compared with that of the punch, we can, however, give no definite rule, for whereas the evils more or less incidental to all

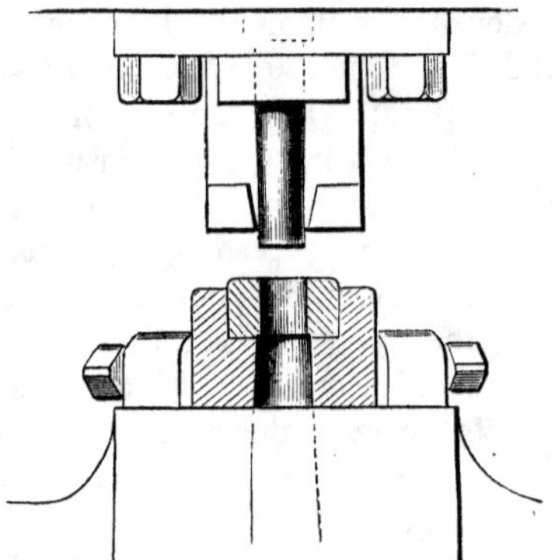

FIG. 193.—Section of Punch and Die.

punched holes—namely, their form being conical instead of cylindrical, and their sides ragged instead of smooth—become greater with every proportionate increase in the size of the die, there appears, on the other hand, to be no valid reason against reducing the clearance to the lowest possible limit consistent with the free passage of the punch, nor for doubting that the effect would be a corresponding reduction in these evils. Ordinarily the diameter of the aperture in the die is made to exceed by one-sixteenth part the diameter of the punch ; but, when it is required to give a more conical form to the holes produced, this difference is increased. But in the generality of holes for rivets, and for most ordinary purposes, the conical form is a very great disadvantage, although cases do occur in which it is even beneficial—as, for instance, in the attachment of two plates only by means of rivets, when the hole in each plate have been punched outwards, each hole being in this way roughly countersunk. Another objection to punched holes is the rounding of their upper edges, which takes place on account of the compression of the iron before the punch begins to tear apart its fibres. This is especially the case with thick plates, and for this, even if for no other reason, they are much less fit subjects for punching than those which are thin. It is possible, however, that if the speed of the punch could be very largely increased, great improvement in this respect and also in the general smoothness of the holes might be the result. With respect to the punch itself—of which the practice is to use it only upon plates of which the thickness does not exceed its own diameter—the very slightly tapered form, shown in Fig. 193, is almost universally given, in order to diminish its friction against the sides of the hole. Inclining its face slightly (at from 3° to 5°), instead of making it horizontal, is by no means so prevalent a practice, though it has the advantage of bringing the pressure more gradually upon the machine than is the case when the work is commenced at all parts of the circumference at the same moment. That the intensity of this pressure is very considerable, even in punching holes of moderate size, may be gathered from the fact that it is found to be very nearly equal to the tensile strength of a bar, the area of whose cross section is the same as the area of the sides of the hole. Thus, in punching a hole 1 inch in diameter through a wrought-iron plate 1 inch in thickness, the pressure required was slightly more than 77 tons, which divided by 3·14 (the area of the sides of the hole)=24·6 tons per square inch, which corresponds very closely with the ultimate tensile strength of wrought iron of average quality. In some experiments by Mr. Hick of Bolton, this rule was found approximately to hold good, even when the diameter of the punch was increased to 8 inches, and the thickness of the plate to $3\frac{1}{2}$ inches, the pressure then required being 2,000 tons; and that it applies as well to shearing as to punching, provided that the edges of the knives are parallel, is also borne out by the facts given by Mr. Anderson in a recently published volume of this series.[1]

In spite of its tapered form, a punch shows much reluctance to quit the hole it has made in any plate or bar, preferring to lift it during its own ascent, unless the weight be excessive. This tendency is resisted by affixing to the upper part of the jaw a 'guard' or 'puller-off,' of which the lower face is at least as low as the face of the punch when it is in its highest position. This puller-off is generally of wrought iron, although a portion of the cast-iron framing is occasionally made to serve the purpose; but, in either case, it must be of sufficient strength to perform the office from which it derives its name. Its position can be seen in each of the machines of which engravings have been given.

With regard to the knives of shearing machines, no very general rule seems to obtain—at least as far as the angle at which each knife is sharpened is concerned. In most instances the lower knife is ground to the same angle as the upper; but in others its edge is made square, that of the upper one being slightly acute. The actual angle which gives the best results varies with the hardness of the material, 75° being about the minimum in the case of wrought iron and steel. Much greater uniformity of practice exists in the angle at which one edge of a pair of shear knives is inclined to the other, by which they are made to

[1] *The Strength of Materials and Structures*, by John Anderson, C.E., &c.

bring a gradual instead of a sudden strain upon the machine. This angle is generally from 3° to 8°, according to the length of the blades, the whole of the obliquity being given to the upper knife, whether the construction of the machine demands that the motion should be imparted to that or to the lower one, of which the edge is horizontal. The lower knife thus forms a guide for the plate which is to be cut; for, by letting it rest fairly upon the edge, the workman is enabled without difficulty to keep it level in the direction of the cut. To assist him in also keeping it approximately level in the opposite direction, an adjustable finger, attached to the upper part of the framing, has been applied. It may be seen on the left-hand side of the machine represented in Fig. 190. With this it becomes impossible for the plate to be suddenly tilted up when the pressure comes upon it, or for the workman to raise its outer extremity to such an extent as to cause it to become jammed between the knives instead of being cut, from which breakage of the knives and damage to the machine occasionally result, and of which there is always some danger if the contact between the edges of the knives be imperfect.

For shearing the edges of boiler and other plates with greater speed and precision than is possible with the preceding machines, in which long cuts can only be produced by combining a number of short ones—placing the plate sufficiently correctly to ensure each one of the series forming part of the same straight line being no easy task—special machines have been constructed. In these the length of the knives is made to exceed or at least to equal the length of the plate, which amounts in some instances to 10 feet and upwards. With these machines a single cut can be taken from the whole length of the plate, and in point of straightness it is of course far superior to any producible in those previously mentioned, in which 18 or 20 inches is the maximum length of the knives. The other limiting dimensions of the shearing and punching machines above figured are as follows :—

The hydraulic-bear (Fig. 186) is capable of punching a hole 1¼ inch in diameter through a 1-inch plate, provided that its centre is not more than 1¾ inch from the edge. With the hydraulic shears, bars of any length up to either 2 inches square, or 4 inches wide by ¾ of an inch thick, can be cut.

The lever machine (Fig. 189) is adapted for punching holes 1 inch in diameter through ⅞-inch plate at any

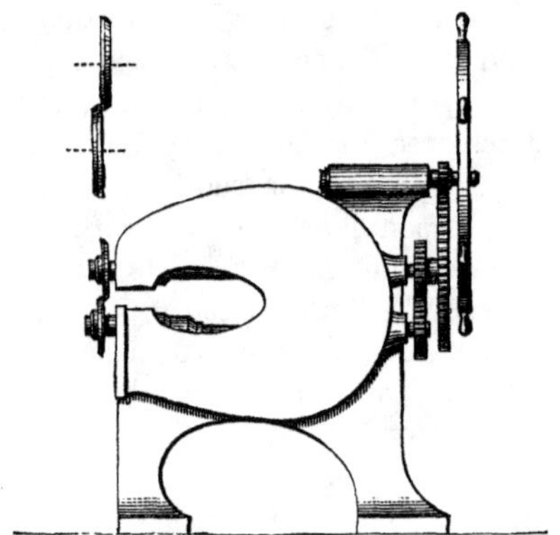

FIG. 194.—Circular Shears.

distance not exceeding 18½ inches from the edge, and for shearing ⅞-inch plates up to 17¾ inches from the edge.

Fig. 190, the double-lever machine, may be used for punching 1¼-inch holes in ¾ inch plate, and for shearing plates ⅝ inch in thickness—the greatest practicable distance from the edge being 12 inches in each case.

The eccentric machine (Fig. 192) may be used for either punching or shearing plates 1¼ inch in thickness at a distance of 18 inches from their edges, provided that the punch does not exceed 1¼ inch in diameter.

One other kind of shearing machine should be mentioned, although it is not often to be found in engineers' workshops. We allude to the *Circular Shears*, invented by Mr. Roberts of Manchester, of which Fig. 194 is an example. The knives are made from circular discs of steel—as shown in the left-hand corner of the figure—and being driven in opposite directions, they can be used for straight cuts of unlimited length, and also for curved work—which frequently gives them the advantage over ordinary shears. In the manufacture of tubes, and for other purposes for which long lengths of sheet metal are required, they are very largely used, and they are also occasionally made for cutting plates up to as much as an inch in thickness. Where large quantities of sheet metal have to be cut into strips, a number of these revolving shears are mounted on one pair of spindles, so that a sheet in being passed between them is cut up at a single operation, the distance between the adjacent pairs of cutters regulating the width of the strips.

CHAPTER X.

ON THE DISTRIBUTION OF THE MOTIVE POWER TO MACHINE-TOOLS.

OUR tour round the workshop must now be brought to an end. It has been but a hasty one—too hasty indeed for us even to become thoroughly acquainted with the cutting-tools which belong to the Fitting-shop, although it is to these that we have been obliged chiefly to confine our attention. But in taking leave of them there not unnaturally arises the question, Whence comes the power which enables them to perform their arduous duties, and how is it conveyed to them and kept under control? These points we will endeavour to clear up as briefly as possible.

With regard, then, to the first of them we may answer, that in this country our coal is the almost invariable source

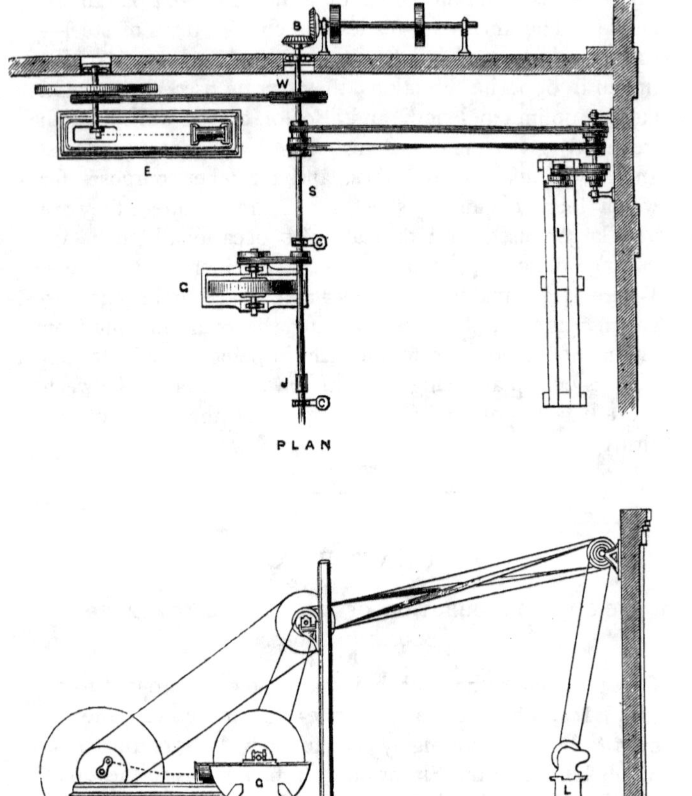

FIG. 195.—Part Plan and Section of an Engineer's Workshop.

of the power, and that steam generated by its means is so applied in a steam-engine as to produce the requisite rotatory motion. And with respect to the distribution of the power throughout the workshop, and rendering it available for driving each individual machine, we have only to point to the long lines of shafting over our heads, together with the countershafts, pulleys, and belts which are there suspended in bewildering confusion. These constitute the greater and lesser arteries of the system, and the instant the life-giving stream ceases to flow through them the whole of the members are left powerless.

Our few remaining pages shall therefore be devoted to examining these somewhat in detail; but it must not be imagined that no other method is employed for transporting to the required points the power obtained from the forces of Nature, any more than that the combustion of coal, &c., is the only one of these forces which can be made available. Had we but the opportunity of extending our observations to the arrangements of the Smith's Shop, and Foundry, we might indeed find instances in which we could speak of the 'stream' of power in a much more literal sense. The steam hammer—that tractable giant for which we are so greatly indebted to Mr. James Nasmyth—would furnish us with a case in point, the steam which works it being conducted in pipes directly from the boiler to an engine which forms part of the machine itself. Hydraulic cranes, again, are worked by an actual stream of water brought under great pressure from an 'accumulator,' and this system, very largely applied by Sir William Armstrong, appears to have a great future before it. But with neither of these can we now concern ourselves.

Some insight into the ordinary method of—as it were—laying on the power by the shafting, &c., just mentioned, may be gained from the diagram, Fig. 195, upon the opposite page. These represent in plan and in section a small portion only of a large workshop, in which it has to be transmitted from the engine E to the lathe L, the grindstone G, &c. The first step is to set up a *main line of shafting*, S, in any convenient position where the wall of the building, or the columns which support the roof (which are marked C, C, in the present diagram) offer facilities for supporting it at suitable intervals. The strength of the shafting must, of course, be in proportion to the strains to which it will be subjected—and on this point the reader will do well to consult another volume of this series [1]—these being due, not only to the torsional stress caused by the transmission of the power, but also to the transverse pressure brought upon it by the *belting*, by which it receives its motion from the engine, and hands it on to the tools, either directly, or through the intervention of *countershafts*, of which we shall have more to say presently.

Where the length of the workshop is great—and it is

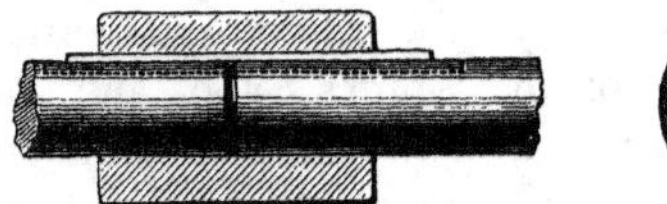

FIG. 196.—Box Coupling ($\frac{1}{12}$).

generally desirable that it should be—the main shaft cannot be conveniently made in a single piece. Fig. 196 shows the best and simplest form of coupling—known as the *box coupling*—by which a series of comparatively short pieces of shafting can be connected together in such cases. The box or ring fits freely over the two ends, and is secured by a taper steel key driven tightly into a bed formed partly in the shafting and partly in the box. It will be seen that this arrangement enables any one length to be readily removed, if required, without disturbing its neighbours. For a joint of this kind to be efficient the coupling should always be placed near a bearing, as shown in the foregoing plan.

Occasionally, however, it is necessary to carry a line of

[1] *The Strength of Materials and Structures*, by John Anderson, C.E.

shafting across some open space—as, for instance, from a fitting-shop to a foundry—where it cannot conveniently be provided with intermediate support, so that the shaft must be made in one continuous length. In such cases much additional strength may be given to it by the system of *trussing*, shown in Fig. 197, a series of tie rods being so arranged that each in turn is thrown into tension as it is brought to the lowest point by the revolution of the shaft.

Fig. 197.—Trussed Shaft ($\frac{1}{60}$).

In the case of horizontal shafts (and the trussed shaft shown in the figure, as well as all the others which we shall have to speak of here, are horizontal), the bearing portions upon which they revolve are commonly called *journals*. These in general — when at least the shafting is 'bright' or turned throughout — do not differ from other portions of it either in diameter or in truth of surface. When, however, the main portion of a shaft is left rough or 'black,' the turning down which is necessary at the bearing reduces the diameter of the journal, as shown at A, Fig. 198. The effect of this—when at least the length of the reduced portion does not exceed the width of the bearing which supports it—is to prevent any endlong motion on the part of the shaft, which of course is not re-

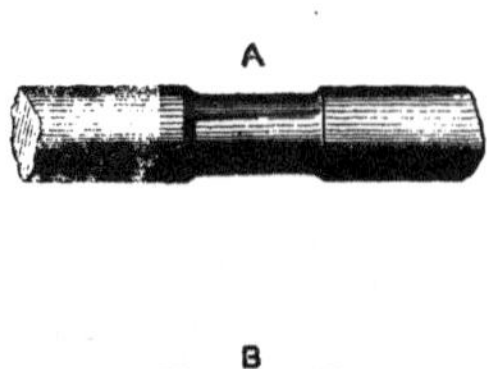

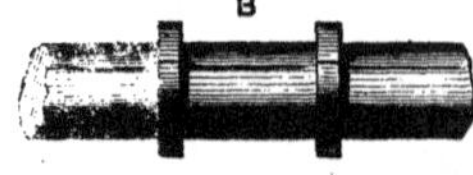

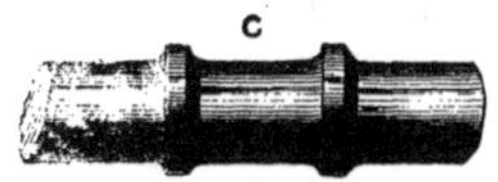

Fig. 198.—Journals.

quired. But in a long length of shafting some slight motion in this direction must be possible, since the expansion or contraction which takes place with every variation in the temperature would cause undue pressure against one shoulder or the other at the points of support. It is, therefore, necessary to allow some liberty at all except one of these points, the travel at this one being prevented in the case of bright shafting by collars, either simply shrunk on as at B, or—which is much better—forged out of the

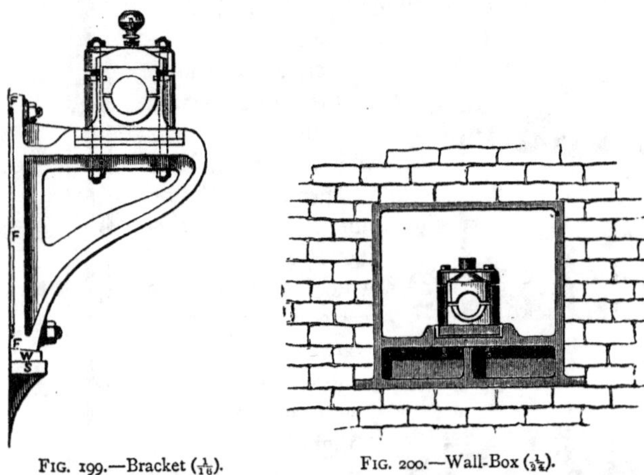

FIG. 199.—Bracket ($\frac{1}{16}$). FIG. 200.—Wall-Box ($\frac{1}{24}$).

substance of the shaft itself, and turned with the internal angles rounded, as at C. The form which this journal frequently assumes when it occurs at the end of a line of shafting will be found in Fig. 204, one of the collars in this case being extended in length to form a boss, upon which either a drum or toothed wheel may be keyed.

Our shafting is now coupled up, and at its extremity or at some one point in its length it is provided with a journal resembling one or other of those just mentioned. We have now to support it either from the cast-iron columns, or the wall of the workshop, or, perhaps, from the beams of the roof.

In either of the two former cases we make use of a cast-iron bracket, such as Fig. 199, varying its form according to fancy, or to suit any special case. For giving a firm footing to this, our column should have upon it a projecting snug (S), on the upper surface of which a wedge (W) enables us to adjust its height with accuracy. Two bolts only, which pass completely through the column, are then required for holding the bracket perfectly firm, *chipping fillets* (F, F, F) running across the back of it, providing us with a ready means of fitting it to the irregularities of the cast-iron surface. In a similar manner a bracket can be attached to a wall, a piece of hard wood packing being then fitted to the much greater inequalities of the brickwork, and the bolts which are passed through it having large cast-iron discs or *wall washers* under their heads for the purpose of distributing the pressure.

In our engravings of the countershafts, Figs. 206 and 207, will be found examples of a *pendent bracket*, for the support of a shaft from an overhead beam, and of a *wall bracket.*

In its course, however, a main line of shafting not unfrequently encounters a wall running in a different direction to its own, as at W, in the plan at the commencement of the present chapter. For supporting it under these conditions the arrangement known as a *Wall Box* is required. This—which is little more than a rectangular cast-iron frame of suitable depth—is secured by being built into the wall, as shown in the illustration on the opposite page, (Fig. 200).

The shaft, however, does not rest directly upon any part of either the wall-box or the bracket. In each case a *Plummer Block* is provided for its reception, of which the object, when used either for the shafting of a workshop or for any of the verv numerous other revolving shafts for which it is employed, is, first, to diminish as much as possible the friction between the moving surfaces: secondly, to transfer the wear from the journal to the more easily renewable 'brasses' in which it runs; and thirdly, to afford the means of adjusting these as the wear takes place. In Fig. 201, a front and also a side view of a plummer block is represented, together with a portion of a *wall plate* upon which it may be conveniently fixed. With the exception of the upper and lower brasses B and B′ and the four bolts (two of which may be dispensed with by the arrangement shown in Fig. 199), the whole of the plummer block is of cast iron, its lateral adjustment being provided for by keys driven in between the extremities of its base and the snugs upon the wall plates. Similarly the height can be varied by increasing or

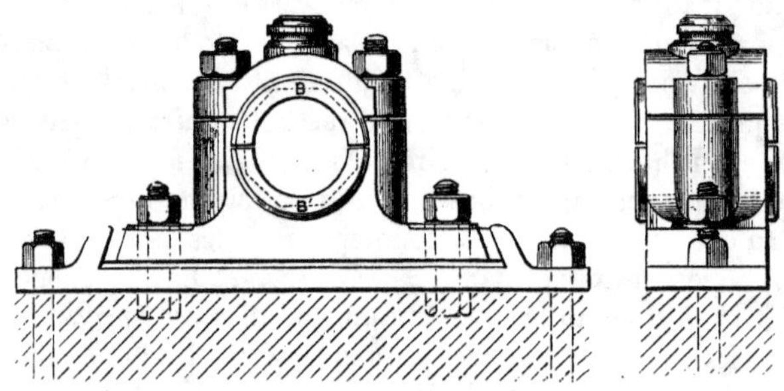

FIG. 201.—Plummer Block (⅛).

reducing the thickness of wood packing below it. As the brasses become worn they can also be adjusted, the lower one by introducing thin 'liners' of tin plate, the upper by bringing down upon it the cast-iron cap; their inclination to revolve with the shafting being resisted by giving an octagonal form to the opening in the plummer block, or by other convenient means. It must not, however, be supposed that the 'brasses' are necessarily made of brass: gun-metal is a better and much more usual material for them; and the friction may be still further reduced by lining them with Babbitts or some similar 'white metal,' which is very largely employed for engine bearings and for other work of a more highly-finished character than that with which we are now dealing.

But the ease with which the journals of a shaft revolve in

their bearings depends as much upon the quality and sufficiency of the lubricating material with which they are supplied, as upon the metal of which each is composed; and this brings us to a subject of such importance in connection with machinery of every kind, that we shall be amply warranted in digressing for a moment to consider it.

First as to the lubricants themselves, with respect to which it may be accepted as an axiom, that the fluidity should be less or more in proportion as the pressure between the surfaces to which they are to be applied is heavy or light. For their effect appears to be to prevent any pair of surfaces from coming into actual contact, by interposing between them a thin film of their own substance. But whenever the lubricant is too thin or the pressure too heavy, this film is pressed out, and some portions at least of the surfaces are brought into contact, adhesion and consequent 'cutting' or 'scoring' of the surfaces being the inevitable result. The fact of the adhesion of clean polished surfaces is well known; —two pieces of plate glass, for instance, when once they have been brought into absolute contact by pressing out the film of air between them, being not only incapable of sliding upon one another but even of being forced asunder; and a pair of accurately made external and internal cylindrical gauges furnishing a similar instance. When these are of equal diameter the external will not pass through the internal gauge, if both be perfectly clean and dry, and, if pressed into it, cohesion sometimes takes place between their surfaces, and no available amount of force will separate them. But a single drop of fine oil, placed either upon the external or the internal gauge, enables the former to pass easily and smoothly, the force of the capillary attraction apparently compressing the steel to a sufficient extent to enable the minute film of oil to insinuate itself everywhere between the surfaces.

The effect of adding a drop of oil to a pair of gauges, of which the external is of slightly smaller diameter than the internal one, may also be mentioned, as, although at first sight paradoxical, it proves the necessity of lubrication under opposite conditions. With a difference between them of only one 10,000th of an inch, these, when dry, pass each other so loosely as to appear not to fit at all. But a drop of fine oil, applied as before, causes them to fit together with moderate tightness, and by using a thicker oil they would undoubtedly appear to fit more tightly still

Besides being of proper consistency, however, it is important that lubricating oil should not be readily liable to change on exposure to the air. For the following list, in which the principal oils which are adapted for the general purposes of engineering workshops are placed in order of merit, we are indebted to Sir William A. Rose, of Upper Thames Street :—

1. Sperm oil.—The best which can be used for general machinery, more especially when accurately fitted and running at a high speed or under a light pressure. It is to be regretted that its greater cost frequently causes inferior oils to be substituted for it.
2. Trotter oil.—For general machinery.
3. Rape oil.—For heavy machinery. Not to be recommended for high speeds or light pressures.
4. 'Journal oil.'
5. Lard oil.—A tolerably good substitute for sperm oil for light machinery.
6. Neatsfoot oil.—Mixed with tallow this may advantageously be employed for lubricating steam pistons and other moderately-heated machinery.
7. Olive oil.
8. Palm oil.—A compound of this with animal fats is used for railway axles under the name of Palm railway grease.
9. Seal oil; also Whale oil.

Next, as to the best methods of applying the oil; a problem by no means easy of solution in the case which we have been considering—namely, that of a line of shafting running in a series of bearings which are often in somewhat inaccessible positions. Some kind of apparatus, capable of delivering a small and constant supply, therefore becomes essential, so as to guard as much as possible against the power being uselessly expended in heating the bearings—which is the certain result of allowing the supply of lubricant to fail. A simple but not very effective plan is to cast a cup at the top of each plummer block, a small oil-hole running from the bottom of this through the cap and the upper brass to the journal. Such a cup appears in Fig. 201. In it is placed a piece of sponge, which readily sucks up the oil poured into the cup, and delivers it up again but slowly through the oil-hole.

As an improvement upon this, the *Syphon Lubricator* was introduced, in which the opening of the oil-hole is no longer at the bottom of the cup, but is brought almost or altogether up to a level with the top of it, by fixing into the aperture a short piece of tube, or otherwise. An ordinary cotton wick, of which one end is passed down the tube and oil-hole, and the other hangs over into the cup, then draws the oil over slowly and continuously, and delivers it into the bearing. For marine engines, and other machinery in which the motion is continuous for long periods, this method answers admirably; but for shafting, &c., which is running only for a portion of each day, the waste of oil during the idle hours is a great objection.

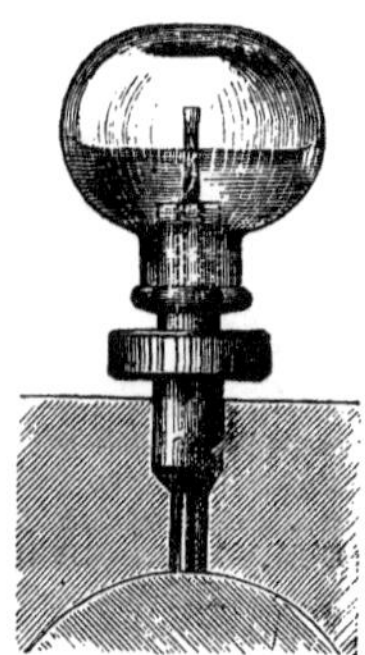

FIG. 202.—Needle Lubricator (⅓).

A much more ingenious arrangement is that known as Lieuvain's *Needle Lubricator*, one form of which is represented in Fig. 202. The oil is placed in a strong glass bottle, the neck of which is then stopped with a wooden plug. Through this plug passes the 'needle'—a piece of iron wire—fitting very loosely into the hole in the plug, which is bushed with brass. The orifice of the oil-hole in the bearing being enlarged to admit the outer end of the plug, the entire lubricator is then turned over into the position shown in the present engraving, and also in that of the bracket and plummer block (Fig. 199). The needle at once drops down upon the journal, the rotation of which imparts to it a slight but continual vibration, which causes the oil to creep slowly down into the bearing as long as the motion continues, but stops the supply directly it is brought to rest. A brass box with a glass 'skylight' is sometimes substituted for the bottle, but for shop shafting the latter is preferable, as the quantity of oil contained in it can be more readily perceived.

Returning now to the general question of the distribution of the power, let us examine the arrangement for drawing off from the main stream the successive portions of it required for serving the various tools in its neighbourhood. For

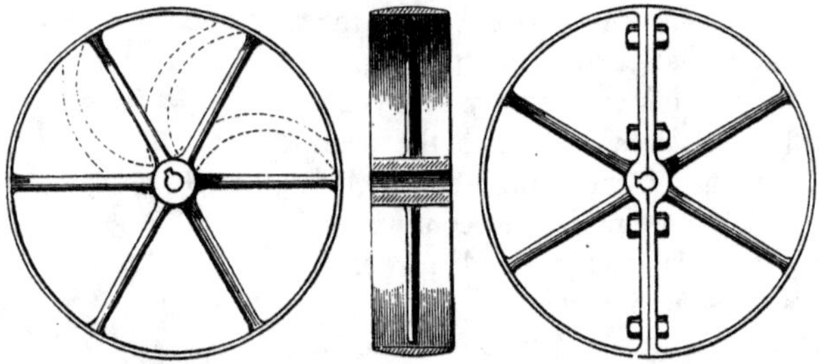

FIG. 203.—Driving Drums.

many of them it suffices to place opposite to them upon the main shaft a *Driving Drum*, such as that shown above, in the left-hand portion of Fig. 203. A leather belt, running round the periphery of this and round a similar drum upon the machine which is to be driven, enables the power to be delivered with but trifling loss at any point within about 30 feet of the line of shafting. In the plan, Fig. 195, the grindstone G is driven in this way—'direct' as it is called—thus receiving from the main shafting whatever power it may require, just as that in turn obtained it from the crank-shaft of the engine. The reader is now too familiar with the system of 'keying' to be for a moment at a loss how to attach the drum securely to the shaft: we will, therefore, only mention, in connection with the first of the two represented in the accompanying diagram, Fig. 203, that in order to diminish the risk of fracture in casting, to which the edges of such thin castings are liable, the arms are frequently curved, as indicated by the dotted lines, instead of being made radial. The other drum shown in the figure is known as a *Split drum*, and it possesses this advantage, that it can be attached at any point in a long line of shafting without necessitating its removal. By so proportioning the size of the aperture through the boss that the act of bolting together the two halves of the drum causes it to clip the shaft sufficiently firmly, the key may often be dispensed with. The slightly convex form of the face which appears in the section might perhaps be thought to be ill adapted for preventing a belt from running off at the edges. For reasons which will be found elsewhere in this series,[1] it is found, however, to be better than any other, and it is the form almost invariably given to a 'single' drum, or one upon which one belt only is to run without change of position. The width of a drum of this kind should be about one-fourth greater than that of its belt.

When the duties of a belt and drum are exceptionally heavy, as in driving a long line of shafting, or when it is necessary to affix to it a toothed wheel for the same purpose, it is advisable to enlarge the shaft at the point of attachment, and to have recourse to keying in the ordinary way, as shown at A, Fig. 204. If requisite, other unsplit drums can still be introduced upon the shaft, the method of 'staking' shown at B enabling them to be secured concentrically, although their bosses must of course have been bored out to a considerably larger size in order to pass over the enlargement.

For the belting, by which the last stage of the connection

[1] *The Elements of Mechanism*: Professor Goodeve, p. 9.

between the prime mover and the several machine-tools is established, various materials—such as canvas and india-rubber, gutta percha, Manilla and other vegetable fibres—have been employed; but for this purpose it may be safely asserted that there is nothing like good leather, to which, therefore, we will confine ourselves.

The size of a leather belt must of course depend upon the amount of power which has to be transmitted by it, or rather by its 'leading side,' for one half of it only—namely, that which is moving towards the driving-shaft—is in action at once, and it is along this only, and not along the 'following side,' that the power is, as it were, flowing. The greatest tension (or pull) which can be brought upon this side by the

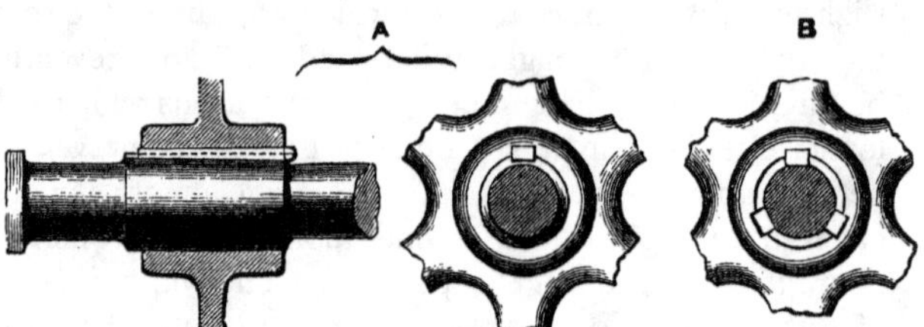

FIG. 204.—Methods of Keying and Staking.

action of the machine, together with that upon the following side due to the stretching of the belt over the drums, furnishes us with a guide as to the sectional area of leather required in each case; 40 lbs. per inch in width for each eighth of an inch in thickness being about the greatest strain at which it may be worked with safety. But as the thickness must depend upon that of the hides from which the belting is cut, and this varies only from about $\frac{1}{6}$th to $\frac{1}{4}$th of an inch, it becomes little more than a question of what width is to be employed, and whether the belt shall be single or double. Examples of the joints used in building up both of these kinds, from any requisite number of short strips, are given in the diagram (Fig. 205), for the strips, being cut straight, cannot, of course, exceed the length of the hide which yields them. Another kind—'ridged belting'—is also there represented, the object of the ridges being to prevent undue stretching of the edges, a defect to which wide single belts are liable.

We have now the means of conveying the requisite power to our machine-tools, and as far as the speed and the direction at which they are driven are concerned, we can vary the first within considerable limits by making the driving-drum upon the shafting larger than that upon the machine (or *vice versâ*), and the second by either crossing

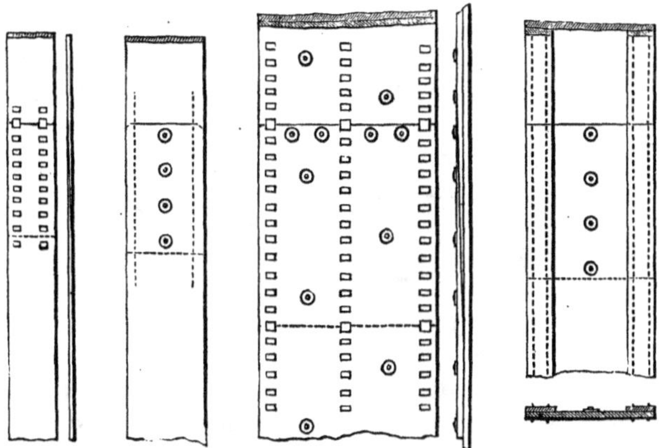

FIG. 205.—Joints of Leather Belting (⅛).

or not crossing the belt in its passage from one to the other. With the simple addition of a 'loose' or 'idle' pulley, to enable the tool to be brought to a stand without stopping the main shafting or throwing off the strap, this indeed is sufficient in the case of many machines, as, for instance, those for punching and shearing; but when it is required either to have the means of readily adjusting the speed or changing the direction, an intermediate or *counter-shaft* becomes necessary. An example of one of these in its simplest form—ready adjustment of the speed only being its object—will be found in the drilling machine represented in Fig. 161, where it is attached to the lower part of the framing. They are however more usually placed overhead, as in the case of the lathe in Fig. 195, since belts cannot be worked with advantage when the distance between two shafts is very small. But when once we are provided with a countershaft, then, by merely increasing the width of its loose pulley (L), and introducing another one, (L′), in the manner shown in the annexed diagram (Fig. 206)—a second drum, or one of greater width being also placed upon the main shaft—we are enabled to drive it at will, either in the same or the reverse direction. For by passing over the loose pulleys two separate belts, one of which is crossed and the other not crossed, they are made to revolve constantly in opposite

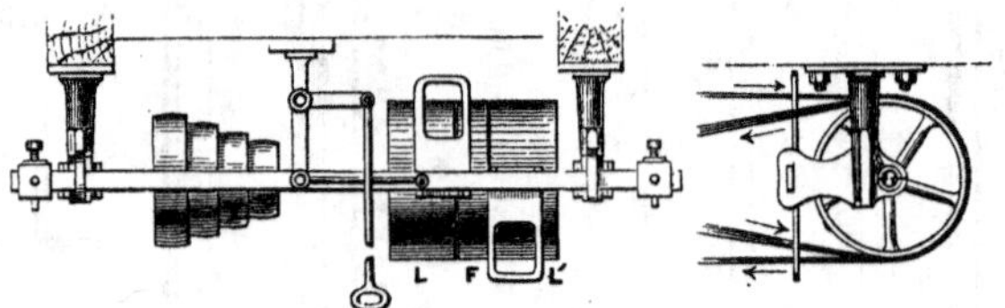

FIG. 206.—Countershaft for Reversing ($\frac{1}{32}$).

directions, and although neither of these is capable of imparting motion to the countershaft, it can be instantly thrown into action by transferring one or other of the belts to the narrow fast pulley between them. With this object each belt is made to pass through a 'fork' just before it reaches its pulley, and these being attached to a sliding bar, which by sufficiently obvious means can be moved either to one side or the other, enable the countershaft—and with it the lathe or other instrument driven by it—to be set in motion either backwards or forwards, or to be brought to rest by the mere movement of a handle.

A slight further alteration renders the speed unequal in the two directions, and gives to the countershaft the means of providing a screw-cutting lathe, or any other tool, with a 'quick return.' The pulleys which give the backward or return motion are then made of much smaller diameter than the others, as in Fig. 207, whilst the size of the drum on the shafting from which the crossed belt passes to them is correspondingly increased. In the diagram the two fast pulleys are marked F and F′, one of each size being now required, and the loose pulleys are indicated by the letters L and L′. Whether each of these is made in two parts, as in the figure, or in one, as shown in Fig. 206, is only a question of convenience in manufacture.

The determination of the relative diameters of the drums and pulleys which will be required for driving any particular machine presents no difficulty, provided we know, in the first place, the speed at which the main shafting revolves, and secondly, that which it is most advantageous to impart

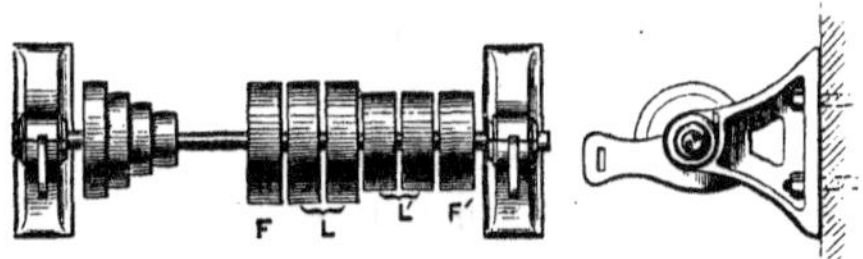

FIG. 207.—Countershaft for Quick Return ($\frac{1}{32}$).

to the tool or the work. Their actual sizes are limited, on the one hand by the circumstance that when the driving drum is very large its weight becomes objectionable, and on the other, that when a pulley is very small, its surface is insufficient to prevent the belt from slipping, unless, at least, the speed be very high, or the power transmitted by it be inconsiderable.

The former of these—namely, the speed of the shafting—is assumed by engineers' tool makers to be 100 revolutions in a minute, and the driving apparatus ordinarily supplied with any machine-tool is therefore so proportioned as to give such an increase or reduction of this speed as shall be suitable for the work to be performed by it. Cone pulleys, to which attention has been already called on more than one occasion, are also introduced whenever it is necessary to have the power of varying the rate of speed, and that this is necessary in the case of such instruments as the lathe will be easily imagined. For not only does the variable diameter of the work produce great differences in the length of it upon which the tool is made to act in any given space of time, but the different degrees of hardness of the materials which have to be treated make it exceedingly desirable to have the means of varying the motion accordingly. The following are the approximate speeds, expressed in lineal feet per minute, at which metal work generally is most advantageously turned or planed :—Steel, 12 ft. ; cast iron, 18 ft. ; brass, 20 ft. ; wrought iron, 24 ft. Dividing these by the number of feet in the circumference of any piece of work which is to be turned, we of course obtain the number of revolutions at which it is desirable that the mandril should be driven. With wood the surface speed per minute may be sometimes as high as 2,000 feet (and even 3,000 feet per minute when the tool instead of the work receives the motion), and it is probable that those given above for metal might be much increased but for the great heat produced, which endangers the temper of the tool.

In order to assist in carrying off the heat, and also to reduce the friction between the face of the tool and the newly-cut shaving, copious lubrication is always desirable in working steel or wrought iron in the lathe or planing-machine, although for cast iron and brass, of which the shavings break up instead of curling, the same necessity does not exist, and these are treated in a dry state. The ordinary lubricant in such cases is a mixture of soap and water, or a weak solution of common (carbonate of) soda may be used, since this likewise restrains the rusting action of the water. But coarse kinds of oil, such as whale and seal oil, or mineral oils from petroleum, &c., are preferred for some purposes,—amongst which may be mentioned screw cutting by means of taps and dies.

Returning for a few moments to the plan in Fig. 195, it should be noticed that the main shafting does not always run in a single straight line throughout the workshop, nor even in two or more parallel lines, although this last is by no means an uncommon arrangement. At the top of the figure is a cross-shaft, at right angles to the main shaft (S), from which the motion has to be imparted to it. As there shown, this is done by means of a pair of bevil wheels (B), but when a small amount of power only is to be transferred from one shaft to another running in a different direction, a leather belt may be employed, in the manner explained in Professor Goodeve's work.[1] For this, however, there must be considerable difference of level between the shafts.

It is generally desirable to have the power of disconnecting some portion of a line of shafting in a more rapid manner than by taking off a coupling. This can be done by means of what are known as *Clutches*. These often consist of merely a pair of circular boxes, with corresponding projections cast upon their adjacent faces. These are placed upon the shafting at the point of junction between two of its lengths, and are so arranged that one can be slid either into or out of contact with the other. The shock with which the revolving box engages with that which is at rest constitutes the chief objection to them, and this is completely obviated by the forms represented in Figs. 208 and 209. In the first of these, the *Cone Clutch*, the connection is established by pressing one half, which has an external conical surface, into the other, which is hollowed and coned

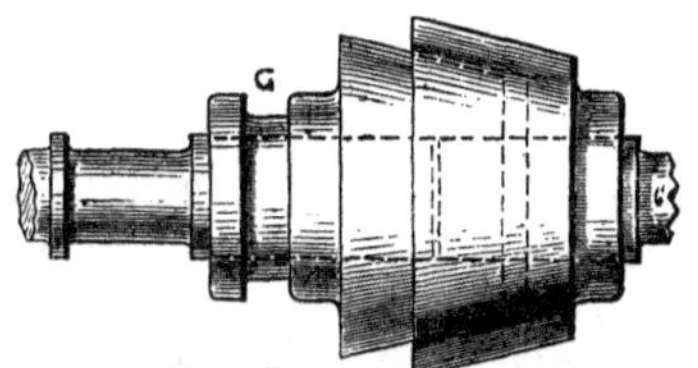

FIG. 208.—Cone Clutch.

[1] *Elements of Mechanism*, p. 11.

internally, the friction between these surfaces being sufficient to cause them to revolve together. In this there is a slight waste of power, on account of the friction of the collar which runs in the groove G, by which, through the medium of a lever attached to it, the sliding cone is continually being pressed into the hollow whilst the clutch is in action.

Weston's Friction Clutch is of more recent introduction, and is said to possess great advantages. The surfaces from which the friction is obtained are much larger in area, being afforded by the flat sides of a series of discs, which are made alternately of iron and elm-wood. In the section of this Clutch, Fig. 209, the latter are marked *a*, and revolve with the shaft A, whilst the former are marked *b*, and are keyed to the shaft B. As here arranged, the clutch is kept always in gear by a series of spiral springs, which are drawn back, when the connection is to be broken, by a lever acting upon the collar *c*.

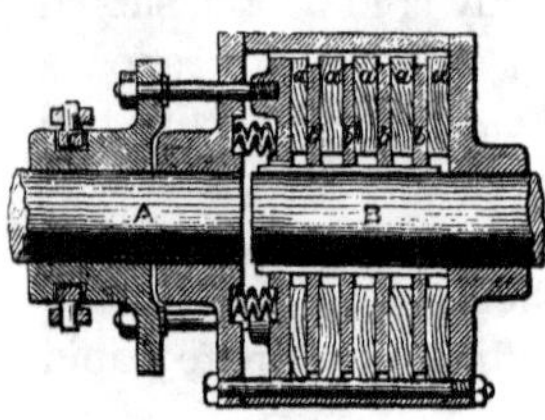

FIG. 209.
Weston's Friction Clutch ($\frac{1}{32}$).

The points in the main shafting at which clutches can be most advantageously introduced, and by consequence the extent to which the wear and tear, and the absorption of power in such portions as would otherwise be 'running idle,' may be saved—for no system of transmitting power with which we are acquainted is entirely free from leakage—must depend upon the general arrangement of the workshop and the disposition of the tools which it contains. For these, however, no absolute rules can be laid down, since each case must in great measure be governed by its own special requirements. To give to the subject the best possible attention in the first instance, will be found in the long run to be the only course which is consistent with true economy.

TABLE OF CHANGE WHEELS.

8-inch Lathe (Centre 16¾) ; *Screw,* 3 *Threads per Inch.*

The small figures over the numbers denote grooves in Radial Arm or Tangent Plate. The two numbers in one column denote double studs or pinions.

Threads Per inch	Left Hand Screws				Right Hand Screws				
	Spindle	Stud	Stud	Screw	Spindle	Stud	Stud	Stud	Screw
1	30	^{2}80	^{2}20, 60	30	30	^{1}35	^{2}40	^{2}20, 60	30
1½	30	^{2}75	^{2}20, 60	45	30	^{1}40	^{2}30	^{2}20, 60	45
2	30	^{1}60	^{2}70	20	30		^{2}120		20
2½	30	^{1}50	^{2}70	25	30		^{2}120		25
3	30	^{1}50	^{2}65	30	30		^{2}110		30
3½	30	^{1}50	^{2}65	35	30		^{2}110		35
4	30	^{1}50	^{2}60	40	30		^{2}110		40
4½	30	^{1}50	^{2}60	45	30		^{2}110		45
5	30	^{1}45	^{2}60	50	30		^{2}110		50
5½	30	^{1}45	^{2}60	55	30		^{2}100		55
6	30	^{1}45	^{2}55	60	30		^{2}100		60
6½	30	^{1}60	^{2}45	65	30		^{2}100		65
7	30	^{1}50	^{2}45	70	30		^{2}90		70
8	30	^{1}50	^{2}45	80	30		^{2}90		80
9	30	^{2}50	^{1}40	90	30		^{2}80		90
10	30	^{2}50	^{1}35	100	30		^{2}80		100
11	30	^{2}50	^{1}27	110	30		^{2}70		110
12	30	^{2}50	^{1}27	120	30		^{2}70		120
13	30	^{2}60	^{1}50, 25	65	30			^{1}65, 60	120
14	30	^{2}65	^{1}35, 20	80	30			^{1}70, 60	120
15	30	^{2}60	^{1}50, 25	75	30			^{1}75, 60	120
16	30	^{2}60	^{1}40, 20	80	30			^{1}80, 60	120
18	30	^{2}60	^{1}45, 20	80	30			^{1}90, 60	120
20	30	^{2}60	^{1}50, 20	80	30			^{1}100, 60	120
24	30	^{2}50	^{1}60, 20	80	30	^{2}35	^{1}21	^{1}60, 20	80

INDEX.

www.ingramcontent.com/pod-product-compliance
Lightning Source LLC
LaVergne TN
LVHW020228110826
845151LV00003B/853

* 9 7 8 1 4 2 5 5 3 2 8 3 3 *